Globalization and the City

John Rennie Short
and
Yeong-Hyun Kim

LONGMAN

Addison Wesley Longman Limited
Edinburgh Gate, Harlow
Essex CM20 2JE
England

and Associated Companies throughout the World

Published in the United States of America
by Addison Wesley Longman Inc., New York

Visit Addison Wesley Longman on the World Wide Web at:
http://www.awl-he.com

This edition published 1999

ISBN 0 582 36912 6

British Library Cataloguing in Publication Data
Short, John R.
    Globalization and the city/John Rennie Short and Yeong-Hyun
    Kim.
        p. cm.
    Includes bibliographical references and index.
    ISBN 0-582-36912-6 (pbk.)
    1. Urbanization. 2. International economic relations.
    3. Cultural relations. 4. International relations. I. Kim, Yeong-
    Hyun. II. Title.
    HT361.S52   1999
    307.76-dc21   98-49367
                                                  CIP

Library of Congress Cataloging-in-Publication Data
A catalog record for this book is available from the Library of
Congress

Typeset by 3 in Sabon and News Gothic.

Produced by Addison Wesley Longman, Singapore (Pte) Ltd.
Printed in Singapore

For **John I. Short** (1933–1997)
and
Wol-Hee Lee

Contents

List of illustrations

*All photographs by John Rennie Short

List of tables

Acknowledgements

We are grateful to the following for permission to reproduce copyright material:

Figure 2.1 'The Future of the State', *The Economist Newspaper Limited*, 20 Sept., 1997; Table 2.1 'Stateless monies, global financial integration and national economic, autonomy: the end of geography?' in *Money, Power and Space* (Martin *et. al*, 1994), Blackwell Publishers Ltd; Figures 9.1, 9.2 and 9.3, City images of Rochester, Los Angeles, CA and Atlanta, Georgia from Hot Spots, *Forbes* (1997).

Whilst every effort has been made to trace the owners of copyright material, in a few cases this has proved impossible and we take this opportunity to offer our apologies to any copyright holders whose rights we may have unwittingly infringed. Any further acknowledgements will be added to a reprint.

PART ONE

Global Discourses

Chapter 1:   Going Global

1

Going Global

Every era has concepts that capture the public imagination, and globalization has recently emerged as one for our time.

(Hall and Tarrow, 1998: B4)

The concept of globalization is an obvious object for ideological suspicion because, like modernization, an earlier and related concept, it appears to justify the spread of Western culture and of capitalist society by suggesting that there are forces operating beyond human control that are transforming the world.

(Waters, 1995: 3)

The postmodernists' claim that we have seen the death of metanarratives is only half-right. The gods of the first half of the twentieth century have fallen from grace and prominence and there is no longer the same belief in Progress and its attendant political philosophies such as Marxism. But the claim is half-wrong because at the dawn of the third millennium, a new 'big story' has taken hold: globalism.

Globalism has two interrelated parts. First, there is the process of globalization which can be broadly defined as the stretching of similar economic, cultural and political activities across the globe. There is now a taken for granted assumption that most of the world is affected by accelerating, widening and deepening processes of globalization. Second, there is the discourse of globalism. Globalization has become a much used term in media, business, financial and intellectual circles. 'Global' has become a common adjective and a ubiquitous shorthand notation in both academic studies and popular accounts of the contemporary world.

In this chapter, we examine these two parts of the story in more detail. We outline the main processes of globalization, explore the discourses of globalism and lay the foundation of a new narrative for globalization.

Globalization

Capital flows across the globe have markedly increased; a vast array of cultural products from different countries have become available in one place; and the nation-state is no longer the only entity that affects people's political life and ideas. Economy, culture and polity are being transformed, reshaped and reworked to produce a more global world and a heightened global consciousness. We can usefully distinguish three related aspects of globalization: economic globalization (a global economy), cultural globalization (a global culture) and political globalization (a global polity). These three elements are interwoven one with another; however, their analytical separation allows us to unpack the complexity of the process into more manageable forms. Here, we will outline the main structure of our arguments while leaving the detail for subsequent chapters.

ECONOMIC GLOBALIZATION

It has been suggested that the world economy has been globalized through the formulation of global production, global markets and more currently global finance. The transnational operations of multinational firms have given rise to a new international division of labor, shifting manufacturing sectors from developed to developing economies. The worldwide production and market expansion of multinational firms has also led to the explosive growth of producer services including financial, legal, consultancy, accountant and advertising firms. Multinationals have significantly contributed to the surge of international trade among diverse parts of the world over the past few decades.

While the global production and market of multinationals are important to the integration of national economies across the world, globalization is most developed in the sphere of finance, creating ever-freer flows of capital on a global scale. The recent rise of global capital flows has been aided by deregulation and liberalization measures which have lowered barriers to international capital movements. Direct investment and portfolio investment crossing national boundaries have increased dramatically. This surge of financial flows on a transnational scale has given rise to an 'increasingly globalizing and intensely competitive world economy' (Swyngedouw: 1996: 1499).

CULTURAL GLOBALIZATION

The increasing movements of people, goods, capital and information have globalized the world. Culture, including ethnicity, language and religion, has traditionally been associated with certain places. Such examples are Japanese in Japan, Hispanics in Latin America and Moslems in the Middle East. As global cultural flows have increased over the past few decades, it has become difficult to find purely territorialized cultural forms. It is not surprising any more to meet Japanese-speaking Americans, Spanish-speaking Latin Americans living in New York City, and East Asian Moslems making a living and a culture in towns in northern England. Cultural traits have been deterritorialized.

Is this global culture dominated by a deterritorialized American culture? Brand names of many American products, such as Coca-Cola, Nike, Burger King, Apple Mac and Holiday Inn, can easily be found throughout the world. American TV programs, including news, talk shows and soap operas, are shown in many cities

outside the US. Hollywood films are dominant in world film markets. It is possible to travel across the world, eating and sleeping only in overseas franchises of American restaurants and hotels. The prevalence of American cultural products around the globe has been considered as crucial evidence of American cultural hegemony in the contemporary world.

Although Americanization, homogenization and commodification are very useful concepts to understand cultural trends in the contemporary world, they highlight only certain aspects of cultural globalization. A deterritorialized American culture is not consumed uniformly around the world. People in the Middle East may understand and interpret McDonald's in more or less different ways from those in Latin America or North America. The growth of cultural flows has increased sameness between distant places; but it has also fostered the complexity of local cultures. The culturally globalizing world is a complex process of the creation of deterritorialized cultures which are reterritorialized in different forms in different places. Global culture is a combination of multiple reterritorialized cultures, rather than a unified culture reflecting American cultural hegemony.

POLITICAL GLOBALIZATION

As economic globalization has proceeded, multinational firms and banks have increased their significance in the world economy, and the nation-state is no longer the predominant institution controlling transactions crossing borders. Multiple market-driven players are now competing with national governments over the rules and practices of international trade, exchange and negotiation.

States, in their various forms, have become more entrepreneurial. While competing and negotiating with other actors in the world economy, national, state and city governments are more market-friendly than managerial and control-oriented. Compared to even just a few decades ago, today's governments throughout the world are more inclined to cooperate with private businesses in the making of tax and employment policy as well as in technology development.

States have become another player in the global market rather than separate control institutions.

International organizations, such as the IMF, the World Bank and the WTO (the GATT before 1995), have also been playing a very important role in prompting political globalization. These supra-national organizations have long served as major institutions promoting the idea of more open international trade and freer markets. They have introduced various structural adjustment policies and imposed market liberalization codes on many national governments, particularly in the developing world. With the Asian financial crisis and resultant IMF bailout packages, many states in the developing economy have no other choice but to follow the IMF's 'advice' in their economic policies on international trade and foreign investment. The crisis of sovereignty in an IMF-era Asia may not be directly linked to the decline of nation-states, but it is suggestive of the shifting role of national governments in the countries where strong state intervention used to be conventional wisdom in the development of the national economy.

Globalization has changed and continues to change the world. Global economic, cultural and political flows have increased dramatically over the past decade. There are, however, some doubts about the nature of globalization. Has globalization made the contemporary world completely different from its past? Or is it merely a phase of a dynamic capitalism which has gone through numerous mutations since the eighteenth century? Perhaps intellectuals exaggerate the impacts of globalization on the world? We should be wary of jumping to the conclusion that today's world is defined exclusively as a global economy, culture and polity. The global economy appears to exclude many parts of the world; global cultural flows are hampered by the persistence of national consciousness; and international organizations still negotiate with state governments rather than master them. We are still a long way off from a global world.

A new metanarrative

Despite the qualifiers we have just made, issues, identities, problems and solutions have been increasingly defined in global rather than national terms. There has been a rise of global consciousness (Robertson, 1990, 1992; Strange, 1995). The term 'global' has become a common adjective in academic studies and popular accounts of the contemporary world (Dicken, 1998; Dunning, 1993; Robertson, 1992; Sassen, 1991; Waters, 1995), though not without voices of dissent (Gordon, 1988; Harvey, 1995; Hirst and Thompson, 1992, 1996; Mander and Goldsmith, 1996). This rhetoric has been reinforced by technology and new representations of the world. Cosgrove (1994) notes that seeing the Earth as a whole through the Apollo space photographs is critical to the totalizing discourses of one-world and whole-earth to which people around the world have become so closely attached.

The growing use of the term 'global' can be easily found in our everyday life: TV advertisements like IBM's 'solutions for a small planet', and 'recognizing Citibank's commitment and capabilities around the world'; a cover story of a magazine, 'Asian software goes global'; and the title of a newspaper article, 'Beijing critic's exile: bow to global pressure' all convey a message of an interconnected, integrated world. In the academy, an immensely increasing body of work has argued that 'Much of what happens in other spaces – communities of various types including, certainly, urban regional, societal, intrastate and supra-state spaces, collectively referred to as local spaces – cannot be fully understood without reference to the global' (Sklair, 1998: 196).

Discourses have political consequences. The discourse of globalism often portrays the process as a force beyond political control. The quote from Waters that began this chapter provides an important warning. The discourse of globalism is often used to attack the legitimacy of traditional leftist politics. In advanced capitalist countries the specter of a global economy is often used to discipline workers to accept lower wages, increased workloads and changing labor practices. Globalization, downsizing and restructuring are terms used not only to describe but also to justify and legitimize changes in capital–labor relations. Harvey (1995: 8) notes, 'In my more cynical moments I find myself thinking that it was the financial press that conned us all into believing in globalization as something new when it was nothing more than a promotional gimmick to make the best of a necessary adjustment in the system of international finance.'

There are important silences of what we may term the global discourse emanating from finance capital. These include any assessment of the redistributional consequences (who gains and who loses), and any discussion of the possibilities that globalism can be managed by specific societies and particular communities. Rather globalism is presented as a force beyond civil society, traditional politics and community control. Globalization is used to justify the destruction of social norms, political traditions and community values. It is presented as a force beyond local control and national discussion. It is used to disenfranchise and delegitimize.

Around the world, but especially in the rich countries, the Left has been fighting against the dominant discourse of globalization, seeing it as a destroyer of hard-won gains. The leftist fear is that workers' rights and wages will be pushed down to the global low. A more upbeat view is sometimes found in countries recently emerging into the global economy. In China, for example, there is tremendous enthusiasm for global connection with all the possibilities of improved wages and the possibility of leverage being used to improve human rights. Like the process itself, the global discourse varies around the world.

Two broad positions can be identified. The optimistic view is that globalization is the improver of everything from wages to political freedoms and environmental standards. The alternative, more pessimistic, perspective sees globalization as the destroyer of everything from social welfare programs to living standards. The optimists argue that globalizing national economies of developing countries is a win–win situation, because foreign products and investment increase economic efficiency, political transparency and overall competitiveness. The pessimists point out the sobering reality of increased polarization within a society and between countries, and the widening gap between the few who are benefiting and the many who are not.

In this book we want to move beyond the simple rhetoric and construct a new narrative of globalization that is politically sensitive to the problems as well as the opportunities offered by globalization. In the rest of this chapter we will sketch the outlines of this narrative. As revealed in phrases such as 'intellectual fascination with globalization and its consequences' (McGrew, 1992) and an 'obsession with globalization' (Walker, 1996), globalization is the trendiest issue in current social science. In this book we want to cast a critical eye over the object of this fascination and obsession. We propose the beginnings of a more critical narrative, structured by the following three concerns.

THE UNEVEN NATURE OF GLOBALIZATION

Globalization is an uneven phenomenon. Its impact varies over space, through time, by social strata and by aspects of our life. We need to conceptualize globalization less as a wave sweeping all before it, and more like a leopard-spot pattern, with small islands of wealth and global connectivity interspersed with marginalized areas and populations. Marginalization may be occurring as much as globalization. The popular discourses of globalism have exaggerated the islands rather than the seas of poverty and marginalization.

The uneven nature of globalization occurs in three ways: geographical unevenness, social unevenness and sectoral unevenness. The uneven process of globalization in geographical terms can be observed in the disparity between developed and developing economies, booming and declining regions, and world and non-world cities. There have always been winners and losers in the regional development of capitalism, but now the inequality between winning and losing places (countries, regions and cities) is aggravated. Places which are more global have a much better chance to take advantage of globalization processes, while less global places are, relatively and sometimes absolutely, losing ground.

The benefits and costs of globalization vary through society. In some societies there is a more even spread of costs and benefits. Exemplars would include the Scandinavian democracies. Elsewhere islands of wealth coexist in seas of poverty (Figure 1.1). Many developing countries in

Figure 1.1 Contrasting street scenes in a globalizing Delhi. (*Photographs by John Rennie Short*)

Africa and Latin America have extremes of wealth and poverty often coexisting in the same city. Despite its strong economy in a globalizing era, the US is revealing a gradually widening gap between the rich and the poor.

Globalization has redistributional consequences between countries and between populations in the same country. Skilled manual workers in traditional industries in the advanced capitalist world have seen a deterioration in living standards. The (broadly defined) middle class in the USA have seen a stagnation in their real incomes over the past 20 years, while key workers in globalized industries such as finance, sport and entertainment now have remuneration that even King Midas would envy. Hollywood stars and major league baseball players have incomes that have grown tremendously, while high school teachers have seen little increase in their living standards in the past 20 years.

Globalization also varies by economic, cultural and political arenas. It is apparent that the globalization thesis has focused more on economic processes, marginalizing cultural and, particularly, political globalization. We thus have more knowledge of economic globalization than of cultural and political globalization, and most knowledge is strongly oriented to developed countries. More case studies on specific places around the world would activate more detailed discussion and balance our understanding of the current state of globalization.

In this book we will directly address the uneven nature of globalization. We will draw upon examples from around the world and look at all three aspects of globalization. Most case studies have focused on economic restructuring and have too often ignored cultural and political globalization. They are important in their own right: cultural globalization affects everyday life in cities, including shopping, diet, fashion and structure of feelings, while the impact of political consciousness can be found in a broad array of urban political agendas. Together the varying degrees of economic, cultural and political globalization structure the distinctiveness of particular cities and shape the arena of urban life. We will look at the range of globalization processes and draw upon examples from around the world to show the connected and unconnected, the costs and benefits, winners and losers.

CENTERS AND MARGINS

Studies of globalization have been biased towards Western core countries, while studies of urban consequences of globalization have focused more on world cities in these countries than on cities elsewhere. Our understanding of the processes of globalization is skewed towards the center. Yet a center implies a margin. In this book we will outline the experience of the margins as well as the center. Our data will draw upon the lower levels of the urban hierarchy, the urban experience in poorer countries and what is happening in the smaller cities.

There has also been a strong Anglo-American bias; for example, among world cities, there has been very little discussion of Paris, and many cities in Asia and Africa in particular have received scant attention (Table 1.1). We have many studies of the experience of London and New York, but much less understanding of what is happening in Tokyo, Shanghai, Accra or Kinshasa. The globalization literature is still parochial and the urban writings are still narrowly Anglo-American in focus.

A DIRTY LITTLE SECRET

Globalization, like all new and exciting topics, has attracted a great deal of attention. Much of it is speculative, hazy and vague. This is to be expected. The complexities and ambiguities are still being explored and identified. However, because of the postmodern turn in much social science writing, little emphasis has been placed on empirical observations. There are a lot of speculative comments about a globalizing world or about the global–urban connection but fewer examples of careful measurements of globalization. This is not to demand a mindless empiricism but to note that much of the writing is neither challenged nor reinforced by empirical testing. Ideas are asserted more than demonstrated. This is what we have referred to elsewhere as the 'dirty little secret of world cities research' (Short et al., 1996). Part of the problem lies in the availability of data. Good data tend not to be international, and international data tend not

Table 1.1 Case studies of cities in the global context (except London, New York and Tokyo)

Region	Case studies
North America	Baltimore (Levine, 1989)
	Boston (Ganz and Konga, 1989)
	Chicago (Esparza and Krmenec, 1994)
	Dallas/Fort Worth (Hicks and Nivin, 1996)
	Los Angeles (Beauregard, 1991; Soja, 1987)
	Miami (Grosfoguel, 1995; Nijman, 1996)
	Minneapolis (Kaplan and Schwartz, 1996)
	Montreal (Coffey, 1996; Levine, 1989)
	Portland (Harvey, 1996)
	San Francisco (Walker, 1996)
	Toronto (Todd, 1995)
	Washington, DC (Abbott, 1996; Fuller, 1989)
Europe	Frankfurt (Keil and Ronneberger, 1994)
	Randstad (Konings *et al.*, 1992; Batten, 1995; Shachar, 1994)
	Zürich (Hitz *et al.*, 1994)
South America	Buenos Aires (Keeling, 1996)
	San José, Costa Rica (Tardanico and Lungo, 1995)
	Mexico City (Ward, 1995)
	São Paulo (Kowarick and Campanario, 1986)
Asia and Australia	Auckland (Moricz and Murphy, 1997)
	Manila and Bangkok (Berner and Korff, 1995)
	Osaka (Rimmer, 1986; Batten, 1995)
	Sydney (Baum, 1997)

to be urban. In this book we will try to ground much of our discussion empirically.

Globalization and the city

There is an ever-expanding body of work on globalization. However, much of it is pitched at such a stratospheric level that it is the view from outer space. In this book we want to shift the perspective to a more terrestrial level; we want to extend our understanding of *how globalization takes place*. In particular we want to explore the connections between globalization and urbanization. This book explores how processes of economic, cultural and political globalization lead to changes in the city and how urban dynamics rework globalization. Globalization takes place in cities and cities embody and reflect globalization. Global processes lead to changes in the city and cities rework and situate globalization. Contemporary urban dynamics are the spatial expression of globalization, while urban changes reshape and reform the processes of globalization.

Globalization takes place in cities, particularly large metropolises. The emerging global system of production, market, finance, service, telecommunications, culture and politics has become spatially articulated through a global network of cities (Alger, 1990; Friedmann and Wolff, 1982; Knight and Gappert, 1989; Knox and Taylor, 1995; Sassen, 1996). Global economic, cultural and political changes have radical effects in restructuring cities around the world.

A major factor underlying urban change is the increasing tie of cities to global trends (Brotchie *et al.*, 1995; Brunn and Leinbach, 1991; Sassen, 1994). Cities are increasingly open to global influences. Appreciation of economic, cultural and political trends on a global scale thus is a prerequisite for understanding urban changes around the world. Indeed, examining the dynamic interplay of global and local forces in particular cities has been an important development in contemporary urban studies (Table 1.2). Focusing on cities, in turn, allows for a more concrete analysis of global processes, as cities serve as the strategic sites of a globalizing world.

The influences of global 'economic' trends on

Table 1.2 New terms in the literature on globalization and urban changes

Topic	Terms
Global force and discourse	global control (Sassen, 1991)
	global logic (Grosfoguel, 1995; Ohmae, 1995)
	global scan (Leislie, 1995)
	globality; globalism (Robertson, 1992)
	globe talk; global babble (McGrew, 1992)
	McDonaldization (Ritzer, 1993)
	transnational civil society (Braman and Sreberny-Mohammadi, 1996)
Global finance	the end of geography; global choice (O'Brien, 1992)
	'global waltz' of capital (Lee and Schmidt-Marwede, 1993)
	postmodern money; phantom state? (Thrift and Leyshon, 1994)
	global money (Sampson, 1991)
	stateless monies (Martin, 1994)
Global urban hierarchy	global matrices (Smith and Timberlake, 1995b)
	'league tables' of cities (Lever, 1993)
World cities	basing points (Friedmann, 1986)
	neo-Marshallian nodes (Amin and Thrift, 1992)
	global front office (Warf, 1991)
	gateway cities (Drennan, 1992)
	broker (Lyons and Salmon, 1995)
	global competitiveness (Kresl, 1995; Noyelle, 1989)
	institutional thickness (Amin and Thrift, 1994)
	territoriality (Budd, 1995)
	world city-ness (Knox, 1995)
Information and telecommnunications	new information spaces (Warf, 1995)
	network cities (Batten, 1995)
	an international 'around-the-clock' city (Batten, 1995)
	urban electronic spaces; telemediated city (Graham, 1997)
	wired city, intelligent city, telematics city (Hepworth, 1990)
Social polarization	dual city; two cities (Castells, 1989; Mollenkopf and Castells, 1991)
	marginality (Sassen, 1994 and 1995)
	divided cities (Fainstein et al., 1992)
Global/local	going global; global dream (Pryke, 1991; Todd, 1995)
	localness (Persky and Wiewel, 1994)
	local embeddedness (Dicken et al., 1994; Tödtling, 1994)
	relocalization (Mander and Goldsmith, 1996)
	global paradox (Naisbitt, 1994)
	Jihad vs. McWorld (Barber, 1995)

restructuring the urban economy, refashioning the urban space and repositioning a city in the global economy have been most studied, while the relationship between cultural and political globalization and urban changes has barely been investigated. In this section, however, we attempt to provide a general picture of the urban impact of economic, cultural and political globalization, which includes intensified competition between cities to attract mobile capital, emerging global metropolitan cultures and rising entrepreneurial urban politics.

THE CITY AND ECONOMIC GLOBALIZATION

One of the most important changes in the world economy over the last decade has been a dramatic increase in the mobility of capital across the globe. Globalization is associated with the hyper-transferability of capital crossing national boundaries. Mobile capital takes many different forms such as footloose high-tech industries, employment, institutions, events (conventions and spectacles) and tourism. A growing pool of mobile capital can be attracted, with the right mix of incentives and attributes, to particular cities.

The past decade has seen increased competition between cities around the world over these mobile investments. Competition between cities accordingly has been acutely intensified by the growth of multinational enterprises which have wide-ranging geographical perspectives when considering new investments; by city governments taking on an increased role in promoting and marketing themselves in an attempt to attract inward investment; by the emergence of new worldwide economic sectors such as financial and producer services; by competition for international institutions to locate within cities; and by competition for global spectacles such as major sporting events, cultural festivals and trade fairs, which generate considerable economic multiplier effects. Haider (1992) terms the fierce competition between cities in the 1990s 'place wars'.

Alongside the intensified competition between cities at a global scale, spatial restructuring is another urban outcome of economic globalization (Beauregard, 1991, 1994; Graham and Spence, 1997; Machimura, 1992; Pryke, 1991, 1994a;

Zukin, 1992). The emergence of new financial districts and luxurious residential areas has been a consequence of the rapid rise of financial and producer services and the massive influx of mobile capital in large metropolises within developed economies. The expansion of 'world-city functions' has been the major source of urban restructuring in world cities like London, New York and Tokyo, where corporate headquarters, bank head offices and financial firms cluster. The vast inflows of foreign capital have recently provided a stimulus for unprecedented property booms and urban center redevelopment in non-world cities around the world, such as Auckland (Moricz and Murphy, 1997), Buenos Aires (Keeling, 1996), Seoul (Kim and Choe, 1997) and Toronto (Todd, 1995).

THE CITY AND CULTURAL GLOBALIZATION

The rise of transnational cultural flows has enabled large cities to taste diverse cultural samplings from all over the world. Such samplings include art exhibitions, opera events and sports competitions. Similar cultural traits such as the minimalist, modernist look of many coffee-houses are found in cities around the world. The formation of global metropolitan cultures is a significant aspect of urban change. Global metropolitan cultures can be generalized as increasing commonalties between large metropolises around the world including built environments, specific lifestyles, policy instruments and dynamic business atmospheres. The prevalence of postmodern architecture in big cities throughout the world makes them look more similar. Increasing connections between cities across borders, such as sister city projects, have led to the adoption of similar policies in urban management.

Behrman and Rondinelli (1992) argue that globalization puts pressure on cities to develop their specific cultures in ways that attract business, investment and high-tech professionals and that convince their own residents and entrepreneurs to remain. The co-presence of homogenizing and heterogenizing trends might be a better phrase to describe the processes of cultural globalization rather than a binary classification of globalizing/non-globalizing.

THE CITY AND POLITICAL GLOBALIZATION

The worldwide transformation in urban governance towards entrepreneurialism is a significant indication of the presence of politically globalizing forces in the city. The steady rise of market forces has undermined the pre-eminence of national and urban governments in controlling urban economies. The promotion of international competitiveness has come to be the hegemonic economic project for many cities. A neoliberalism now dominates the discourse of urban economic development around the world (Hall and Hubbard, 1996; Harvey, 1989; Peck and Tickell, 1995; Young and Lever, 1997). The concept of world cities has been swiftly taken up by national and urban planners to be used as a policy-generating idea (Machimura, 1998; Shachar, 1994). Enhancing the position of a particular city towards world city status has become a top priority of what we can term 'wannabe' world cities. Todd (1995), for example, examined Toronto's commitment to achieving the status of a world city in terms of the local politics of 'going global'. He noted that the promotion of Toronto's global competitiveness (a freer market, a smaller state, increasingly 'recommodified' delivery of social services) becomes a compelling project that seeks to resolve the more abstract problems of conflicts between particular and general interests by mobilizing political support and appealing to common-sense understandings of the economy. Cox (1995) coined the term 'the new urban politics' in the contexts of globalization and competition and their relation to the politics of local economic development.

'Globalization and the city' is not a new topic in urban studies. It has been examined, but existing studies have commonly revealed a few limitations, including their exclusive focus on large metropolises in developed economies and the urban outcome of economic globalization. Few studies have attempted to measure the impact of globalization and local initiatives. Few researchers have been successful in grounding political globalization.

In this book we seek to expand our understanding of globalization in four distinct ways. First, we review a wide body of literature. The debates have expanded enormously in recent years. This book aims to provide a map of this new terrain. Second, we link space and place, global and local, by concentrating on the complex relationships between processes of globalization and urban change. Much of the existing debate on globalization is stuck at the level of abstract global space. By focusing attention on specific places we seek to deepen our understanding of how, why and when globalization takes place. We pay particular attention to the uneven nature of global flows and their redistributional consequences. We also consider how the processes operate not only from global to local but also from local to global. Our goal is to present material that is both theoretically informed and empirically grounded. Unlike many studies which focus only on globally induced changes to the city, we look at how the processes operate in the opposite direction and we highlight the role of urban dynamics on reproducing globalization. Third, we sharpen the definition of globalization from a vaguely defined general process to a more precise threefold division of economic, cultural and political globalization. Studies of globalization and urban dynamics have focused more on economic globalization than on the other two. Yet only a full consideration of all three can lead to a deeper understanding. The connections between these three elements are a key concern. Fourth, we extend the range of cities beyond London and New York, to smaller cities and to cities outside North America and Europe.

PART TWO

Economic Globalization and the City

2

Economic Globalization

The bourgeoisie cannot exist without constantly revolutionizing the instruments of production, and thereby the relations of production, and with them the whole relations of society ... Constant revolutionizing of production, uninterrupted disturbance of all social relations, everlasting uncertainty and agitation, distinguish the bourgeoisie epoch from all earlier ones. All fixed, fast-frozen relations, with their train of ancient and venerable prejudices and opinions are swept away, all new-formed ones become antiquated before they can ossify. All that is solid melts into air, all that is holy is profaned.

(Marx and Engels, first published 1872, 1968)

The transnational movement of people, goods, capital and information has been a significant feature of the world economy for over 500 years. What makes the globalization of the past 20 years or so distinctive is that there has been a definitive shift in the proportion of the world's economic activity that is transnational. At the same time, there has been a shift in the nature and organization of transnational economic activity, with the global flows of services, capital and information becoming just as important as, and in some cases even more important than, the international trade in raw materials and manufactured goods.

Globalization is most developed in the spheres of finance, services and telecommunications accelerating free capital and information flows across borders. In this chapter we will consider the emergence of these three major forces of economic globalization: the global financial system, the global service economy and the global telecommunications network. We will conclude with a discussion of the limits to economic globalization.

Global financial integration

The globalization of finance refers to the increasing freedom of movement, transfer and tradability of money and finance capital across the globe. Globalization is not simply synonymous with multinational banks and finance houses, or the process of internalization associated with the increasing presence and importance of such multinational companies in domestic financial markets. Globalization combines these elements with a strong degree of integration between different national and multinational parts. Proclaiming 'the end of geography', O'Brien (1992) argues that money in a globalizing world will continue to try to avoid, and will largely succeed in escaping, the confines of the existing geography. As barriers to free capital flow have fallen and continue to fall at the global level, most financial institutions around the world will be linked to the globally integrated system of world financial centers in the next decade.

Martin (1994: 255) identifies three phases in the historical evolution of the financial system of advanced capitalist countries (Table 2.1); they are the regional and bank-oriented, the national and market-oriented, and finally the transnational and securitized form. In the latest phase, 'capital and money markets are separating from industry, money has been commodified, and as national financial centres become increasingly globalized and globally integrated, it is now national monetary autonomy that is being challenged' (Martin, 1994: 255). Lee and Schmidt-Marwede (1993) call the contemporary circuit of financial capital the 'global waltz' of capital that transcends

Table 2.1 The evolution of the financial system

Regional and bank-oriented form	National and market-oriented form	Transnational and securitized form
Associated with industrialization phase of economic development	Characteristic industrial maturity phase of economic development	Associated with post-industrial and transnational phase of economic development
Banks main source of external funds needed by private sector firms	Capital markets main source of funds, using savings of private investors	Bulk of funds obtained through capital and credit markets, using mainly resources of institutional investors
Industrial growth financed by loans, risk capital and profits	Capital markets channel personal and other savings into industry; risk spread across shareholders	Separation of capital and money markets from industry and commodification of money; proliferation of monetary products
Regional and national banking system; local sources of capital important	Concentration and centralization towards national banking and capital markets; loss of regional financial autonomy; emergence of internationalization	Development of globally integrated system of world financial centers; loss of national financial autonomy to supranational economy of stateless monies

Source: Martin (1999: 256)

national boundaries and serves to subject the world economy to an ever more geographically extended but highly integrated and hence competitive process of change and differentiation. Figure 2.1 shows the stunning growth of international transactions in bonds and equities and the daily turnover on the foreign-exchange market over the past decade.

The rise of global capital flows has stemmed from a number of interrelated changes that have been restructuring financial markets in the developed economy since the early 1970s. The changes include the progressive deregulation of financial markets at both the national and international levels, the introduction of an expanding array of new financial instruments and monetary products that allow riskier, bigger and more easily tradable financial investments, the emergence of new market actors, especially institutional investors such as large pension funds, and the spread of new telecommunications and information technologies that have extended and accelerated financial transactions (Cohen, 1996; Corbridge *et al.*, 1994; Herring and Litan, 1995; Lee and Schmidt-Marwede, 1993; O'Brien, 1992; Walter, 1988; Warf, 1995). Of these changes, notes Martin (1994), market deregulation and technological innovation in telecommunications have been particularly influential to global financial integration.

Deregulation of financial markets has been tightly linked to the breakdown of the Bretton Woods System in the 1970s (see the appendix to this chapter) (Corbridge, 1994; Kenen, 1994; Roberts, 1994; Strange, 1994). The collapse of the system led to a new geography of international finance, as stateless monies continued to move offshore, or beyond national regulatory regimes. The formation of the global financial system was aided by new telecommunications which facilitated a global information system which gave instantaneous information on the trend of global markets (Mookerjee and Cash, 1990), electronic funds transfer (Langdale, 1985) and the distribution of financial communities (Code, 1991).

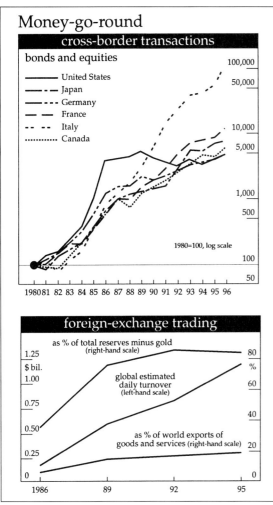

Figure 2.1 Accelerated global capital flows; bonds and equities are gross purchases and sales of securities between residents and non-residents.
(*Source: The Economist, 1997e: S24*)

Global service economy

The growth and diversification of service industries is one of the most important economic changes in the second half of the twentieth century. The shift to services can easily be assessed by the increasing proportion of service industries in total employment and gross domestic product (GDP) in many countries, specifically in developed economies. New telecommunications and information technologies in combination with the emergence of multinational service firms and deregulation have greatly facilitated worldwide transactions in services (Daniels, 1993, 1995).

The globalization of services has been caused mainly by the growth of foreign direct investment (FDI) in services by multinational firms. Services are currently the fastest increasing sector of multinationals' investments in both developed and developing countries, outperforming manufacturing production. The export of managerial skills, information and organizational techniques is a major form of intrafirm cross-border trade of multinationals (Dunning, 1993).

The explosive growth of producer services, such as accounting, advertising, computing and data processing, engineering, financial, legal, and management consulting services, has contributed to the rise of services as the leading sector in the contemporary world. Major trends toward the development of multinational manufacturing, services and financing industries have created an expanded demand for these specialized service activities to manage and control global networks of factories, service outlets and branch offices (Sassen, 1991, 1994). Bagchi-Sen (1997) argues that the internationalization of producer service firms has stemmed from their client-following activities. A large number of US service firms have followed their manufacturing clients who have invested abroad for several decades. Bagchi-Sen also notes that these US service firms have been successful in finding new markets beyond their existing US clientele. Growing pressure from domestic competition and advancements in telecommunication and information technologies have greatly contributed to the global expansion and dominance of US service companies.

The high concentration of these producer services in major world cities like New York, London and Tokyo has received mounting attention (Graham and Spence, 1997; Lyons and Salmon, 1995). The simplest explanation for the concentration is a need for face-to-face communication. Unlike other types of services, according to Sassen (1990, 1991), producer services are not dependent on proximity to the consumer served. Rather, such specialized firms need to locate near other firms that produce key inputs or whose proximity makes possible joint production of certain service offerings. The recent growth of producer services and the expanding global reach of multinational firms have immensely reinforced each other. As highly specialized producer services have increasingly located in a handful of world cities, the importance of these cities in a globalizing economy has increased.

Global telecommunications network

The convergence of information technologies and telecommunications has given rise to communication systems and networks that stretch across the globe. In the past two decades, three major innovations – the fax, the mobile telephone and the Internet – have proved how telecommunications networks can be used to create new mass-market products that change the way people live and work (*The Economist*, 1997d). Seamless technology advances in communication networking have been fostering both the supply and demand of global telecommunication systems and services (Garcia, 1995). Supply and demand at the global level have been stimulated particularly by a series of developments, such as reductions in the cost of service, improvements in networking capability, the privatization and commercialization of the telecommunication sector, and the emergence of global providers of telecommunication and information-based products and services. Given an expanding and ever more integrated global telecommunications network, parallels can be drawn between the role of telecommunications in the current phase of the world economy and the respective roles of railways in the age of industrialization and automobiles in the post-war boom (Gibbs and Leach, 1994).

Revolutionary transformations in telecommunications have been linked to the rise and expansion of a global financial system and a global service economy (Brunn and Leinbach, 1991; Daniels, 1993; Langdale, 1989; Moulaert and Djellal, 1995; Warf, 1989, 1995). The international spread of new telecommunications and information technologies, coupled with progressive deregulation, has facilitated and accelerated international transactions in finance and services.

The formation of the global telecommunications network is well illustrated in the phenomenal growth of the Internet and the World Wide Web, which are truly international information infrastructures and the most important mechanisms for the transmission and distribution of academic knowledge around the world (Figure 2.2). Although the Internet has existed in some form or other for more than 25 years, it has become much more popular in the past few years. More than 35 million host computers now provide services to users around the world (Network Wizards, 1998). Web sites are becoming an outstanding source of all kinds of information, partly replacing libraries, newspapers, marketing catalogs and advertising brochures.

Advanced telecommunications require massive investment in infrastructures, such as teleports, fiber-optic networks and intelligent (smart) buildings, and continuous incorporation of new technologies, discoveries and innovations. Significant investments are concentrated in only a few selected cities where there is an agglomeration of corporate headquarters and financial, legal and advertising services (Daniels, 1991; Langdale, 1991; Moss, 1991; Moulaert and Djellal, 1995; Warf, 1989, 1995). As vital information transmission facilities, such as teleports equipped with satellite earth stations, exhibit high fixed costs and low marginal costs, they offer significant economies of scale to small users unable to afford private systems (Warf, 1995). New telecommunications technologies and their infrastructures, therefore, are currently becoming concentrated in a handful of world cities such as New York, London and Tokyo that contain a pool of demands (Hepworth, 1991; Longcore and Rees, 1996; Moss, 1987; Sassen, 1991; Warf, 1989).

Since the presence of advanced telecommunications systems can progressively promote the growth of future-oriented information-intensive industries, telecommunications systems are also considered important tools for boosting urban economic development (Castells and Hall, 1994; Daniels, 1991; Ducatel and Miles, 1992; Gibbs and Leach, 1994; Hepworth, 1990). The growing focus of current urban policy innovation on telecommunications is termed 'urban telematics policy' (Gibbs and Leach, 1994; Graham, 1994).

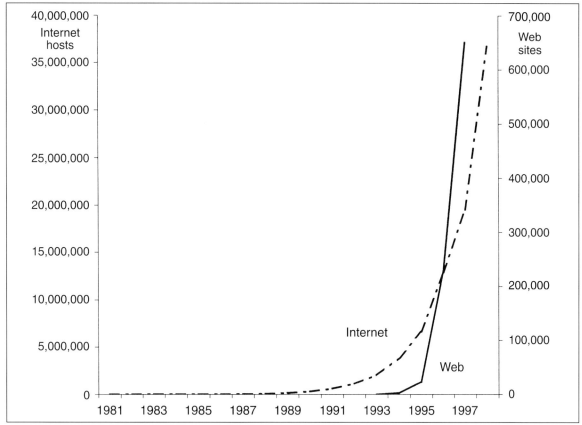

Figure 2.2 The growth of the Internet and the World Wide Web. (*Sources*: Network Wizards, 1998, http://www.nw.com/zone/host-count-history; Gray, 1997, http://www.mit.edu/people/mkgray/net/web-growth-summary.html)

Table 2.2 Characterization of urban places and electronic spaces

Urban places[a]	Urban electronic spaces[b]
overcome time constraints by minimizing space constraints	overcome space constraints by minimizing time constraints
territory	network
fixity	motion or flux
embedded	disembedded
material	immaterial
visible	invisible
tangible	intangible
actual	virtual or abstract
Euclidean or social space	logical space

[a] Based on buildings, streets, roads, and the physical spaces of cities
[b] Constructed 'inside' telematics networks by the use of computer software
Source: Graham (1997: 120)

Graham (1997) even argues that telecommunications-based urban development provides challenges to how urban space is conceptualized and planned within advanced industrial cities (Table 2.2). Although Graham's portrait of telecommunications-mediated urban life in electronic spaces is somewhat exaggerated, the proliferation of computer networks and resultant diminishing spatial constraints is now an established fact of life in contemporary cities of developed economies.

Limits to economic globalization

The shift from an international to a more global economy has become part of the conventional wisdom. However, there are some skeptics who challenge the strong version of the economic globalization thesis. Their objections to the globalization account set the contemporary world economy into its *longue dureé* context. Globalizing tendencies in the world economy, claim skeptics of globalization, have existed in some form or another since the beginning of capitalism, thereby it is largely a myth to think that we are now entering a new era (Hirst and Thompson, 1996). Revolutions in communications and information technology and deregulation in the late twentieth century have further developed, rather than created an economic system.

Contributors to an edited revisionist volume by Kevin Cox (1997) point to the fact that not all firms are footloose, some remain firmly in place, technology transfer between many countries remains problematic, challenges to global production by multinational companies remain significant, and financial services are often not easily transferable. Economic globalization has its limits, as there are still constraints to the emergence of a fully global economy. While adherents of the strong economic globalization thesis, the globalizers, point to the progress to date in the creation of a global economy, the skeptics remind us of the limits.

The role of the nation-state and its economic sovereignty are also important areas of dispute between globalizers and skeptics. While 'a borderless world' has become a glib phrase to describe today's world (Ohmae, 1995), many have tried to specify what a fully globalized world would look like and how different it would be from the current state of the world (Waters, 1995) (Table 2.3). The governed nature of the present world economy has been strongly emphasized by these counter-arguments (*The Economist*, 1997e; The World Bank, 1997).

Compared to an ideal type of a globalized economy, argue Hirst and Thompson (1992, 1996), the contemporary world economy is imperfectly governed through the limited cooperation of the major trade blocs and nation-states, but it is governed nonetheless and in a way that limits the power of transnational corporations and global financial markets. Hirst and Thompson conclude that we do not have a fully globalized economy; what we do have is an international economy and national policy responses to it. Martin (1994) also notes that different financial centers and different national institutions specialize in different circuits of the global capital market, and in this regard global finance still retains a strong national component. Based on a case study on Japan, Carnoy (1993) maintains that nation-states can be key actors in shaping what multinational enterprises do, although multinationals are overpowering economic actors in the international economy and play a key role in shaping the context in which nation-states formulate their own economic strategies. Assessing the comparatively successful performance of the Dutch economy in the past decade, Thrift (1994a) points out the neglect by the globalization literature of the importance of social regulation by the nation-state. Hirst and Thompson (1996) point to the state's exclusive control of territory in the regulation of population. People are less mobile than money, goods or ideas: in a sense they remain 'nationalized', dependent on passports, visas and residence and labor qualifications.

Another objection to the globalization thesis is that the 'global' economy encompasses only a few selected countries that have been increasingly economically interconnected, and that there is no truly global economy. Storper (1992) notes that the global economy may be thought of as consisting of a series of technology districts in the

Table 2.3 An inventory of economic globalization

Dimension	Ideal-typical pattern of globalization	Current state of affairs
Trade	• absolute freedom of exchange between localities • indeterminate flows of services and symbolic commodities	• minimum tariff barriers • substantial non-tariff and cultural barriers • regional neomercantilism
Production	• balance of production activity in any locality determined only by physical/geographical advantages	• international social division of labor being displaced by a technical division of labor • substantial decentralization of production • dematerialization of commodities
Investment	• minimal FDI; displaced by trade and production alliances	• TNCs being displaced by alliances arrangements but considerable FDI remains
Organizational ideology	• flexible responsiveness to global markets	• flexibility paradigm has become orthodox but very substantial sectors of Fordist practice remain
Financial market	• decentralized, instantaneous and 'stateless'	• globalization largely accomplished
Labor market	• free movement of labor • no permanent identification with locality	• increasingly state regulated • considerable individual pressure for opportunities for 'economic' migration

Source: Waters (1995: 94)

advanced economy instead of embracing Third World countries. The 'global' network of fiber-optic cables connects only North America, Europe and northeast Asia, excluding most parts of Africa, Latin America, southeast Asia and the Middle East (The Petroleum Economist and TeleGeography, 1996).

Table 2.4 shows the dominance of developed countries in international trade and foreign direct investment (FDI) flows. Developed countries accounted for more than 65% of world trade in 1995. Alongside developed countries, four Asian newly industrializing countries (NICs) – Hong Kong, Korea, Singapore and Taiwan – and China have actively participated in international trade. The table also indicates that developed countries are the main source of FDI (92.5%) around the world and that they are, at the same time, the main recipients of FDI flows (58.6%). China, Korea and Singapore received about half of FDI in developing countries in 1996 and China, in particular, accounted for 29.4% ($40.2 billion) of investments in developing countries.

The dominance of developed countries is much more evident in the flows of portfolio investment, involving international transactions in corporate

ECONOMIC GLOBALIZATION **21**

Table 2.4 Global distribution of trade and foreign direct investment (FDI)

| | Trade, 1996 ($ bil.) | | FDI, 1996 ($ bil.) | |
	Exports	Imports	Outflows	Inflows
World total	5,249.6 (100.0)	5,369.5 (100.0)	318.1 (100.0)	330.8 (100.0)
Developed countries[a] (A)	3,546.7 (67.6)	3,527.6 (65.7)	294.4 (92.5)	194.0 (58.6)
Developing countries	1,702.9 (32.5)	1,841.9 (34.3)	23.7 (7.5)	136.8 (41.4)
Asian NICs[b] and China (B)	702.4 (13.4)	720.5 (13.4)	11.3 (3.6)	51.9 (15.7)
A + B	4,249.1 (81.0)	4,248.1 (79.1)	305.8 (96.1)	245.9 (74.3)

[a] 22 industrial countries
[b] 4 Asian NICs, but Hong Kong and Taiwan not included in FDI
Sources: International Monetary Fund, 1997, *International Financial Statistics Yearbook*; International Monetary Fund, 1997, *Balance of Payments Statistics Yearbook*, Part 2

securities, bonds, notes, money market instruments and financial derivatives, and other investment including transactions in currency and deposits, loans and trade credits. In 1996, developed countries accounted for 93.5% of the world total portfolio investment assets ($582.7 billion), while they provided 81.5% of other investment assets ($766.8 billion) (International Monetary Fund, 1997). Despite a rapid growth in international trade and FDI, often cited as undeniable evidence of economic globalization, their regional clustering raises strong doubts about the geographic scope of globalization. Developing countries play a very minimal role in the recent surge of global financial flows. The actual scope of economic globalization, as revealed through these data, is much more modest than most commentators suggest.

The *Far Eastern Economic Review* (1996a) compares the contemporary world economy to 'The New Nike Economy: World without Borders'. Nike's Air Max Penny is made up of 52 components collected from five different countries (USA, Taiwan, South Korea, Indonesia and Japan). It is interesting to note, however, that Nike connects only five countries in the Asia-Pacific region to make a pair of basketball shoes, not the entire world. In this case, the globalization of production is certainly exaggerated, although Nike may be sold globally.

A global economy is in the process of becoming rather than being. The globalizers point to the areas of global connectivity, the skeptics highlight the spaces in between. There is a dialectic in which the accelerating pace of globalization highlights the limits to the process. Limits in a double sense: economic and political constraints which undermine the globalizing tendencies through the brute facts of economic geography and the political power of nation-states; and geographic limits to areas and sectors of the world that are globally connected. A global economy is in the process of becoming, but it is not uniform, nor complete, nor all-encompassing.

Appendix: Bretton Woods

The Bretton Woods International Monetary System (1944–1971) emerged in the wake of World War II as a consequence of the economic nationalism that had developed in the 1930s. According to Strange (1994), the 1930s depression convinced the Americans and many Europeans that the world market economy could not be left to work by itself. Rules were needed: exchange rates of different currencies should be made more stable; trade barriers should be lowered. These convictions, in short, were in line with the Keynesian ideas about the role of states and markets. Corbridge (1994) also defines the Bretton Woods agreement as an attempt to grant a good deal of economic sovereignty to individual countries, while also providing for a stable international economy that would foster openness and cooperation between countries.

In 1944, the Bretton Woods agreement was signed, setting up what was essentially a US-run international financial system with three poles – the World Bank, the International Monetary Fund and fixed exchange rates – with the US dollar serving as the convertible medium of currency with a fixed relationship to the price of gold (Thrift, 1989: 34). By the late 1960s, however, the system was under pressure. Countries and companies could not find sufficient international reserves and the US was no longer such a dominant economic power. Fixed exchange rates disappeared and every domestic currency became convertible into every other. Movements of dollars in and out of the offshore markets were adding to the uncertainty and instability of exchange rates. The establishment of a pool of Eurodollars and transnational Eurocurrency lending were crucial steps in a long trend toward deregulating and liberalizing global finance.

3

The Global Urban System

The global hierarchy reflects and embodies social, economic and demographic changes.

(Short, 1996: 37)

The global urban system embodies and connects the global economy. In this chapter we will look at various ways of measuring different positions of cities in the global urban system.

Smith and Timberlake (1995a) identify two dominant methodological strategies in mapping the global urban hierarchy: the attributional and the linkage. The attributional strategy involves ranking cities by the performance of individual cities, e.g., the number of multinational corporate headquarters. The linkage-based strategy concentrates on the interactions between cities to identify a set of nodes in the global urban network. We will look at each in turn.

Command centers of the global urban system

Cities differ in terms of global competitiveness and global connectivity. Ranking global competitiveness and connectivity is extremely valuable in identifying recently emerging urban economic bases and explaining general patterns of winning or losing cities in a crowded global market. There are a number of criteria that can be used for ranking cities around the world. Friedmann (1986) suggests seven indicators: major financial center, headquarters for TNCs, international institutions, rapid growth of business services sector, important manufacturing center, major transportation node and population size (Table 3.1). Some other researchers add new criteria such as telecommuni-

cations (Hepworth, 1990; Warf, 1989, 1995), quality of life (Simon, 1995), international affairs and cultural centrality (Knox, 1995; Rubalcaba-Bermejo and Cuadrado-Roura, 1995) and destination of immigration (Friedmann, 1995). Having good indicators, however, does not guarantee a precise measurement of individual cities' competitiveness in the global economy. Indeed, Friedmann (1995: 40) himself admits that 'The thumbnail sketches are drawn from newspaper accounts and sporadic readings over the last few years and do not reflect original research.' In Friedmann's work, as with many other researchers, world cities are very much pre-defined. Data are found to confirm their world-city status rather than their status being defined by criteria. In a comprehensive critique on world cities research, Short *et al.* (1996) point to the lack of comparable data on cities around the world and refer to this 'dirty little secret' in world city research.

In the global economy, the productivity and competitiveness of cities are to be determined by the extent of the concentration of corporate commanding power, world-class financial institutions and high-order producer services and by the connectedness to international telecommunications and transport networks. Castells (1994) adds informational capacity and quality of life as crucial factors determining a city's position relative to others. In addition, governmental effectiveness, urban strategy, public–private sector cooperation and institutional flexibility can also be of importance in understanding urban competitiveness (Kresl, 1995).

Four of the most frequently used indicators in ranking cities are: command functions, financial

Table 3.1 The global urban hierarchy

1 Global financial articulations

London[a] A (also national articulation)

New York A

Tokyo[a] A (also multinational articulation: SE Asia)

2 Multinational articulations

Miami C (Caribbean, Latin America)

Los Angeles A (Pacific Rim)

Frankfurt C (Western Europe)

Amsterdam C or *Randstad* B

Singapore[a] C (SE Asia)

3 Important national articulations (1989 GDP > $200 billion)

Paris[a] B

Zürich C

Madrid[a] C

Mexico City[a] A

São Paulo A

Seoul[a] A

Sydney B

4 Subnational/regional articulations

Osaka–Kobe (Kansai region) B

San Francisco C

Seattle C

Houston C

Chicago B

Boston C

Vancouver C

Toronto C

Montreal C

Hong Kong (Pearl river delta) B

Milan C

Lyon C

Barcelona C

Munich C

Düsseldorf–Köln–Essen–Dortmund (Rhine–Ruhr region) B

[a] National capital. Major immigration targets are listed in *italics*.
Population (1980s): A 10–20 million; B 5–10 million; C 1–5 million
Source: Friedmann (1995: 24)

markets, producer services and telecommunications infrastructure. We will look at each in turn.

COMMAND FUNCTIONS

Transnational corporations have been identified as the single most important actor in the world economy (Carnoy, 1993; Dicken, 1998; Dunning, 1993). These institutions have played a significant role in creating an integrated, worldwide network of production, exchange, finance and corporate services arranged in a complex hierarchical system of cities (Feagin and Smith, 1987). Between one-fifth and one-quarter of the total world production in the world market economies is produced by TNCs. Total sales for the 500 global industrial corporations in 1996 accounted for 11,435 billion dollars and they employed 35.5 million people (*Fortune*, 4 August 1997).

Since Hymer (1972) noted that the functions of top management in corporate structures concentrate in the world's major cities like New York, London, Paris, Bonn and Tokyo, many world cities researches have used the number of headquarters of the top 500–1,000 transnational corporations as a criterion for positioning cities along the hierarchy (Bosman and Smidt, 1993; Cohen, 1981; Feagin and Smith, 1987; Meijer, 1993; Rosenblat and Pumain, 1993; Short *et al.*, 1996). The distribution of headquarters of major multinationals is definitely indicative of the concentration of economic decision-making power in particular cities. Sassen (1991) argues, however, that the number of corporate headquarters located in a city is a less adequate measure of economic power than it was in the 1960s and 1970s, because producer and financial services are more important factors in the orchestration of the global economy.

Many studies on the global urban hierarchy tend to restrict their geographical scope to Europe (Bosman and Smidt, 1993; Dieleman *et al.*, 1993; Hall, 1993; Lever, 1993; Meijer, 1993; Rosenblat and Pumain, 1993; Rubalcaba-Bermejo and Cuadrado-Roura, 1995; Shachar, 1995). Apart from Short *et al.* (1996) and a few other studies that attempt at setting up a 'global' urban hierarchy, most writers identify the major command centers of urban Europe. There is certainly a

Table 3.2 Locations of the headquarters of the world's largest corporations (ranked by sales)

City, country[a]	1997	1993	1990	1985	1980	1975	1970	1965	1960
Tokyo, Japan	18(5)[b]	17(3)	12(2)	10	6	8(1)	5(1)	2	1
New York, USA	12(1)	6(3)	7(5)	12(5)	10(4)	10(5)	25(8)	29(9)	29(8)
Paris, France	11(1)	3	5	4	7(2)	4(1)	0	0	0
Osaka, Japan	7(3)	4(1)	2(1)	1(1)	1	2	1	0	0
Detroit, USA	4(2)	2(2)	2(2)	2(2)	2(2)	2(2)	3(3)	4(3)	5(2)
London, UK	3(1)	5(2)	7(2)	7(3)	8(3)	8(3)	7(3)	10(3)	7(3)
Chicago, USA	3	4(1)	2	5(2)	4(2)	6(2)	5	6(1)	6(2)
Munich, Germany	3	3(1)	2	1	1	1	1	1	1
Amsterdam, Netherlands	3	0	0	0	0	0	0	0	0
Seoul, South Korea	2(1)	4(1)	2(1)	4	0	0	0	0	0
Frankfurt, Germany	2	2	1	1	1	2	2	2	1
Zürich, Switzerland	2	1	1	0	0	0	0	0	0
San Francisco, USA	1	3	3	1(1)	1(1)	1(1)	1(1)	1(1)	1
Rome, Italy	1	2(1)	2(2)	2(2)	1(1)	1	1	0	0
Düsseldorf, Germany	1	2	2	1	2	1	1	1	1
Courbevoie, France	1	1	3	0	3	3	2	1	1
Fairfield, USA	1(1)	1(1)	1(1)	1(1)	2(1)	1(1)	0	0	0
Cincinnati, USA	1	1	1	1	1	1	1	1	2
The Hague, Netherlands	1(1)	1(1)	1(1)	1(1)	1(1)	1(1)	1(1)	1(1)	1(1)
Stuttgart, Germany	1(1)	1(1)	1(1)	1	1	1	1	1	1
Wolfsburg, Germany	1	1(1)	1(1)	1	1	1	1(1)	1	1
Turin, Italy	1	1	1(1)	1	1(1)	1	1	1	1
Leverkusen, Germany	1	1	1	1	1	1	1	1	1
Ludwigshafen, Germany	1	1	1	1	1	1	1	1	1
Seattle, USA	1	1	1	1	1	1	1	1	1
Vevey, Switzerland	1	1	1	1	1	1	1	1	1
Boulogne-Billancourt, France	1	1	1	1	1	1	1	1	1
Toyoda City, Japan	1(1)	1(1)	1(1)	1(1)	1	1	1	0	0
Essen, Germany	1	1	1	0	2	2	0	2	3
Caracas, Venezuela	1	1	1	0	1(1)	0	0	0	0
Irving, USA	1(1)	1(1)	0	0	0	0	0	0	0
Philadelphia, USA	1	0	0	1	1	2	1	0	0
Madrid, Spain	0	2	2	0	0	0	0	0	0
Los Angeles, USA	0	1	2	4(1)	4(1)	3	5	2	3
Pittsburgh, USA	0	1	2	3	3(1)	4(1)	2(2)	5(1)	8(2)
Milan, Italy	0	1	2	0	1	1	1	1	0
Eindhoven, Netherlands	0	1	1	1	1	1(1)	1(1)	1	1
Atlanta, USA	0	1	1	1	0	0	0	0	0
Basel, Switzerland	0	1	1	0	1	0	0	0	0
Boston, USA	0	1	1	0	0	0	0	0	0

[a] After ranking cities according to the number holding the world's 100 largest corporation headquarters, we trimmed down the list to the top 40 cities

[b] The figure in brackets gives the number of the world's top 20 corporations in that city

Source: Fortune (1961, 1966, 1971, 1976, 1981, 1986, 1991, 1994, 1998)

'European bias' in world cities research originated from easier access to comparable data on European cities and the recent emergence of a more integrated European urban system. In contrast, most large cities in Latin America and Africa have failed to be considered by the global urban hierarchy thesis.

Cities with many corporate headquarters have been called global command centers for business decision-making and corporate strategy formulation (Cohen, 1981; Feagin and Smith, 1987; Rosenblat and Pumain, 1993). Feagin and Smith (1987) note that the headquarters of the largest 500 multinational firms are disproportionately located in large metropolises in developed countries, particularly New York, London and Tokyo. Sassen (1991) also maintains that top-level control and management of the industry has become concentrated in a few leading financial centers, notably New York, London and Tokyo.

We can assess this claim by looking at relevant data. *Fortune* began to report the 500 largest industrial corporations in the USA in 1955, and the 100 largest outside the USA in 1958. The list of the biggest foreign firms was extended to 200 in 1963, to 300 in 1972 and to 500 in 1976. Since 1974 *Fortune* has also published a list of the 50 largest industrial corporations in the world, extended to the 100 largest in 1989 and the 500 largest in 1990. In 1994, the *Fortune* ranking in 1994 began to cover corporations of the service industries (banks, wholesalers, insurance and entertainment, among others) to make the Global 500. We have concentrated our attention on the 100 largest corporations (industrial corporations until 1993; both industrial and service corporations in 1997). Since the yearly lists of corporations outside the USA had no information on the location of corporation headquarters, we obtained these data from *Who Owns Whom* (UK edition in 1966; Continental edition in 1969/70) and *Fortune*'s web sites on the directory of companies. We looked at the location data carefully to ensure that corporation headquarters beyond the narrow administrative boundaries of a city, yet within the wider functioning urban region, were included in a city's total. For example, the New York total includes corporations headquartered in Armonk, White Plains and

Purchase as well as those in the New York metropolitan area, while the Chicago total embraces corporations headquartered in Highland Parks, Northbrook, Glenview and Hoffman Estates.

Table 3.2 shows the urban location of the headquarters of the world's largest corporations. A total of 111 cities have hosted at least one headquarters of the world's 100 largest (industrial) corporations. The time-series data reveal some dramatic changes in the ranking of cities. The most significant is the stunning growth of Tokyo and the decline of the number of headquarters in New York. In 1960 Tokyo had only 1 corporate headquarters and even that was not in the top 20. By 1997, however, Tokyo was the home to 18 of the top 100 corporations and to five of the very largest (see Figure 3.1). In the same period New York declined from 29 of the top 100 and eight of the top 20 in 1960 to only six and three respectively by 1993. The rise of New York between 1993 and 1997 may be due to the change of *Fortune*'s data collection which began to include service corporations in 1994. While New York has been losing commanding power in industrial corporations, it still maintains power in service ones. The table is suggestive of changing global economic fortunes evidenced from the decline of cities in the US, including Chicago, Pittsburgh and Los Angeles, and the growth in Paris, Osaka, Seoul and Munich. In summary, three trends can be noted:

- the growth of Tokyo
- the decline of the older industrial US cities, especially New York
- the increase in a range of cities including Paris, Osaka and Seoul.

FINANCE

There has been an explosion of interest in assessing the urban impact of the formation of the global financial system (Pryke and Lee, 1995). A number of studies have sought to answer the fundamental question, 'Why do financial activities tend to be highly concentrated?'. The geographical clustering of financial institutions, such as bank head offices, stock markets and security firms, has been regarded as the most important factor generating the buoyancy of a handful of world cities in the

Figure 3.1 Two images of Tokyo: (a) as a command center of the global economy, but (b) still a very Japanese city. (*Photographs by John Rennie Short*)

competitive strength in a global financial market. An article on European finance and investment in the *Financial Times* (1990) adds the importance of avoidance of over-regulation in controlling financial businesses. Focusing on the competitive bases of London, Lee and Schmidt-Marwede (1993) point to physical infrastructure as well as the effectiveness of financial markets. Alongside economic factors, socio-cultural aspects of financial centers are emphasized. Budd (1995) argues that the development and maintenance of the territoriality of a city, with global credentials, is at the heart of intense competition between major financial centers. Thrift (1994b) and Amin and Thrift (1994) also stress the social and cultural determinants of international financial centers, namely, 'information, expertise and contacts'.

Reed (1989) ranks New York and London at the apex of the global urban hierarchy as supranational centers (Table 3.4), while Lee and Schmidt-Marwede (1993) and Warf (1989) confirm Sassen's current world economy. Table 3.3 lists the competitive edge that has made emergent world financial centers more attractive to financial institutions than any other cities. Noyelle (1989) identifies the agglomeration of demand, supply, financial intermediaries and business services, technologically innovative environment and quality workers as the major sources of New York's (1991) three cities, London, New York and Tokyo, as global financial centers. There is much agreement on the dominance of London, New York and Tokyo in the hierarchy of financial centers, yet the category of second-tier cities is unclear. Warf (1989) suggests Los Angeles, Toronto, Hong Kong, Singapore, Bahrain, Paris, Zürich and Frankfurt, while Lee and Schmidt-Marwede (1993) place

Table 3.3 Determinants of the competitiveness of a financial center

Author	Determinants
Noyelle (1989)	agglomeration of demand
	agglomeration of supply
	agglomeration of financial intermediaries
	environment of innovation
	technological environment
	availability and cost of labor
	other operating costs
	agglomeration of support business services
Financial Times (1990)	well-run disciplined markets
	multi-currency dealing
	pool of world-class talent
	a central bank
	avoidance of over-regulation
Lee and Schmidt-Marwede (1993)	enabling infrastructure
	built environment of commerce
	size of financial centers
	production of financial centers
Thrift (1994b)	business organizations – sociability and proximity
	markets – large size, rapid dissemination of information and quick response
	culture – information, expertise and contacts

Geneva, Zürich, Boston, Paris, Frankfurt and San Francisco just below the global centers. Reed (1989) identifies Amsterdam, Frankfurt, Paris, Tokyo and Zürich as international centers followed by a number of host centers.

In the existing literature there is a heavy reliance on common assertions of centrality and a limited discussion of the empirical basis of such positions. In order to provoke a more informed discussion, we have selected time-series data for two major indicators of a city's position as a financial center in the global urban hierarchy:

- the number of head offices of the worlds' top 100 banks in each city
- the value and the number of listed companies of major stock markets.

BANKS

Since 1970 *The Banker* has published a yearly list of top banks in the world. The list began with the top 300; it was then extended to 500 in 1980 and to 1,000 in 1989. *The Banker* also provides the location of the head offices of these banks. We have used this source to generate the data in Table 3.5, which shows the head office location of the 100 largest banks from 1969 to 1996. While these data are indicative of world city status, they are not conclusive; but even with this proviso in mind the data are still revealing.

There are 68 cities in which one or more of these head offices have been located during the period of study; of these, only 28 have been the site for two or more head offices, 15 of these have had three or more, and only nine cities have been

Table 3.4 Global financial centers

Hierarchy	Cities	Characteristics
Supranational centers	New York and London	Managers of large amounts of foreign financial assets and liabilities; net suppliers of foreign direct investment capital to the rest of the world; located close to a large number of large industrial corporations; active users of global communication facilities; management meccas that attract and generate ideas and information in ways that eventually establish the organizational and operating norms that govern internationally active organizations
International centers	Amsterdam, Frankfurt, Paris, Tokyo and Zürich	Headquarters for large internationally active banks which influence events that pertain to global asset and liability management
Host centers	Basel, Bombay, Brussels, Chicago, Düsseldorf, Hamburg, Hong Kong, Madrid, Melbourne, Mexico City, Rio de Janeiro, Rome, San Francisco, São Paulo, Singapore, Sydney, Toronto and Vienna	These centers enhance their own financial infrastructures and capabilities by attracting relatively large numbers of foreign financial institutions from a large number of countries

Source: adapted from Reed (1989: 250 and 264)

the location for four or more head offices. Compared to corporate headquarters, the location of bank head offices has been more concentrated in a small number of cities. Tokyo, Paris, Frankfurt, London, New York and Osaka have possessed approximately half of the 100 head offices each year, and most of the largest 20. From the most recent data of 1996 we can identify these six cities as global financial centers with Tokyo clearly dominant. Comparison of the data over time indicates that, in terms of the number of large bank head offices, the importance of Tokyo, Paris and Frankfurt as financial centers has increased, whereas London and New York have declined. In 1969 New York and London combined had nine of the largest 20 banks, but by 1996 this had been reduced to only five. This shift reflects the changing financial strength of the respective national

banking systems. Japanese banks, in particular, have increased their absolute and relative asset strength while the number of American banks belonging to the top 20 as well as the top 100 has decreased. The biggest single change since 1990 is Beijing's growth as a financial center. Economic liberalization and the globalization of the Chinese economy have increased banking activity in the Chinese capital. Although it seems too early to call Beijing a global financial center, the city is certainly increasing its presence in global finance.

Another way to determine the financial importance of world cities is to consider the importance of their banks in overseas markets. Comparative international data, however, are not available. However, data are available for one of the largest markets, the US. *American Banker* (28 February 1995) reported the top 50 foreign banks

Table 3.5 Head office locations of the world's largest banks (ranked by assets)

City, country[a]	1996	1990	1985	1980	1975	1969
Tokyo, Japan	13(5)[b]	16(8)	17(6)	16(4)	14(3)	11(2)
Paris, France	10(2)	9(4)	9(4)	8(4)	5(4)	6(1)
Frankfurt, Germany	6(1)	6(1)	6(1)	5(2)	5(2)	4
London, UK	5(3)	5(2)	5(3)	5(3)	7(2)	6(3)
Beijing, China	5(1)	2	1	0	0	0
New York, USA	4(2)	6(1)	6(2)	7(3)	7(4)	10(6)
Osaka, Japan	4(2)	4(2)	4(2)	4(2)	4(2)	3(2)
Brussels, Belgium	4	2	4	5	4	2
Munich, Germany	3	3	3	4	3	3
Toronto, Canada	3	3	3	2	2	3(1)
Milan, Italy	3	3	2	2	3	3
Amsterdam, Netherlands	2(1)	2(1)	3	3	2	2
Zürich, Switzerland	2(1)	2	2	2	2	2
Rome, Italy	2	2	2	2	2(1)	2(1)
Montreal, Canada	2	2	2	2	2	2(1)
Charlotte, USA	2(1)	1	0	0	0	0
Madrid, Spain	2	0	1	2	3	3
San Francisco, USA	2(1)	2	2(1)	3(1)	3(1)	3(1)
Melbourne, Australia	2	2	1	1	0	0
Stockholm, Sweden	2	2	0	3	3	2
Stuttgart, Germany	2	1	0	0	0	0
Düsseldorf, Germany	1	1	1	2(1)	2	2(1)
Nagoya, Japan	1	1(1)	1(1)	1	1	1
Basel, Switzerland	1	1	1	1	1	1
Siena, Italy	1	1	1	1	1	1
Turin, Italy	1	1	1	1	1	1
Utrecht, Netherlands	1	1	1	1	1	1
Hanover, Germany	1	1	1	1	1	0
Yokohama, Japan	1	1	1	1	0	0
Bilbao, Spain	1	1	0	0	1	0
Chiba City, Japan	1	1	0	0	0	0
Brasilia, Brazil	1	0	1	1	1(1)	1
Columbus, USA	1	0	0	0	0	0
Santander, Spain	1	0	0	0	0	0
Wiesbaden, Germany	1	0	0	0	0	0
Edinburgh, UK	1	1	0	0	0	1
Sydney, Australia	1	1	1	2	2	2
Chicago, USA	1	0	2	2	2	2
Minneapolis, USA	1	0	1	0	2	2
Boston, USA	1	0	1	0	1	1

[a] After ranking cities according to the number holding the world's 100 largest bank head offices, we trimmed down the list to the top 40 cities
[b] The figure in brackets gives the number of the world's top 20 banks in that city
Source: The Banker (1970, 1976, 1981, 1986, 1991, 1997)

Table 3.6 Head offices of the top 50 foreign banks in the US (1994)		
City, country	Assets[a]	Loans[b]
Tokyo, Japan	14	15
Paris, France	6	5
Osaka, Japan	4	4
Toronto, Canada	3	2
London, UK	3	1
Montreal, Canada	2	2
Zürich, Switzerland	2	2
Frankfurt, Germany	2	2
Rome, Italy	2	1
Hong Kong	1	1
Amsterdam, Netherlands	1	1
Utrecht, Netherlands	1	1
Nagoya, Japan	1	1
Dublin, Ireland	1	1
Basel, Switzerland	1	1
Santander, Spain	1	1
Munich, Germany	1	0
Düsseldorf, Germany	1	0
Naples, Italy	1	0
Madrid, Spain	1	0
Turin, Italy	1	0
Tel-Aviv, Israel	0	3
Seoul, South Korea	0	2
Milan, Italy	0	1
Sapporo, Japan	0	1
Beijing, China	0	1
Brasilia, Brazil	0	1
Total	50	50

[a] Top 50 foreign banks in US commercial bank assets
[b] Top 50 foreign banks in commercial and industrial loans
Source: American Banker, 28 February 1995

in the US by commercial and industrial loans and commercial bank assets. It would be more beneficial for this study if this information were provided in a time-series, and if data on American bank offices in foreign countries were also provided. Despite these problems, the data in Table 3.6 give another indication of the relative financial

dominance of selected world cities. The dominance of Tokyo is apparent, although Paris and Osaka are also fairly prominent, followed by Toronto and London. The same data source can also be used to provide an indication of the banking importance of cities in the US. According to *American Banker*, which has provided data on 915 banks as of 30 June 1994, American cities with the most foreign banking offices were New York (408, 44.6%), Los Angeles (116, 12.7%) and Chicago (71, 7.8%).

STOCK MARKETS

The size of the stock market is also a good indicator of a city's financial power and relative position in the global urban hierarchy. The Tokyo Stock Exchange has provided data on major stock exchange markets around the world since 1981. Table 3.7 highlights the prominence of large stock markets in New York, Tokyo and London, compared to Frankfurt and Paris and other cities which are too minimal to list in the table. Among the three big players, the New York stock market is dominant in market value and trading value, reflecting the predominance of the Dow Jones and NASDAQ markets in world stock exchanges. London has more foreign companies listed than the other two global cities, although it was recently exceeded by Frankfurt. Thanks to the Single Market of Europe, London and Frankfurt are more international, while both New York and Tokyo are linked primarily to their domestic markets (Lee and Schmidt-Marwede, 1993). Tokyo's market value had increased dramatically during the 1980s, yet New York has outperformed other cities in the 1990s leading the booming economy of the US.

Overall, the data point to a sharp concentration of financial power in a handful of cities, especially Tokyo followed by New York and London. The data also point to the emergence of Paris and Frankfurt as important nodes in the chain of global capital transactions. They are emerging as future rivals to the big three in terms of both the number of large banks, and the growth of their stock exchanges, which have grown more than ten times in market value over the period 1981 to 1996. Few studies have commented on the recent rise of Paris

Table 3.7 Major stock exchange markets

		New York	London	Tokyo	Frankfurt	Paris
No. of listed	1996	2,617	2,423	1,766	433	686
companies	1990	1,678	1,946	1,627	389	443
(domestic)	1981	1,533	2,659	1,402	197	586
No. of listed	1996	290	533	67	560	187
companies	1990	96	613	125	354	226
(foreign)	1981	37	482	15	173	166
Market value[a]	1996	7,118.4	1,759.4	2,968.2	641.4	591.7
($ bil.)	1990	2,692.1	858.2	2,821.7	341.0	304.4
	1981	1,242.8	204.5	379.7	63.5	53.1
Trading value	1996	4,063.7	1,400.6	929.3	628.2	283.2
($ bil.)	1990	1,325.3	544.1	1,303.1	348.9	120.2
	1981	382.4	36.7	179.8	5.6	12.6

[a] Excludes the figures of foreign companies
Source: Tokyo Stock Exchange, 1982, 1991 and 1997, *Fact Book*

because, we would argue, they have not been empirically informed. The data also allow us to make comparisons with the assertions of other studies. Cities like Chicago, Beirut and Bahrain that are named as global financial centers by O'Brien (1992) do not figure as highly in our data. Since Friedmann (1986), Hong Kong and Singapore have supposedly functioned as major financial centers in Asia (Martin, 1994; O'Brien, 1992; Sassen, 1994) however, this study has indicated that they are not major financial centers. These examples highlight the need for a more careful assessment of the data before making statements pertaining to the global urban hierarchy.

PRODUCER SERVICES

Producer services have grown rapidly in the last decade. There is a growing body of literature on the role of advanced producer services in urban and regional economies. Urban specialization in producer services has been associated with higher median income and fast economic growth in some large metropolises of the US (Drennan, 1992; Drennan *et al.*, 1996). Sassen (1991) calls the city with highly specialized services and financial inno-

vation 'a site for post-industrial production'. It is widely acknowledged that London, New York and Tokyo are distinctive in terms of their disproportionate shares of national and international producer services employment (Bosman and Smidt, 1993; Daniels, 1991, 1993; Meyer, 1991a). Here a question can be raised, 'Do producer services contribute to altering the nature of the global urban system?'.

The rise of producer services has enabled multinational firms and banks to coordinate their branches and offices distributed around the world. Housing a pool of producer services, accordingly, is strongly indicative of a city's economic power in global markets. Meyer (1991a) argues that business intermediaries (producer services) have very important impacts on changes to the world system of metropolis. But his analysis is limited to the long-term processes involved in the rise and fall of some key global cities like Venice, Amsterdam, London and New York, excluding most other cities around the world. Based on an examination of the national urban hierarchies of Japan, the United Kingdom and the United States, Sassen (1991) acknowledges that the locational distribution of producer services follows the accepted

Table 3.8 Cities with large advertising markets[a]

Rank	City	Total local shop billings, 1997 ($ mil.)
1	New York	37,697.2
2	Tokyo	31,449.3
3	London	15,316.8
4	Chicago	11,214.4
5	Paris	9,366.7
6	Los Angeles	7,394.0
7	Detroit	6,760.9
8	San Francisco	5,789.7
9	Minneapolis	5,091.9
10	São Paulo	4,884.6
11	Frankfurt	4,390.4
12	Boston	4,104.4
13	Milan	3,828.4
14	Sydney	3,304.6
15	Düsseldorf	3,271.2
16	Seoul	3,246.1
17	Amsterdam	3,177.3

[a] Local shop billings for the cities outside the US are based on 1,505 agencies
Source: Advertising Age, 27 April 1998
(http://adage.com/dataplace/archives/dp224.html)

urban hierarchy, yet she also strongly suggests that producer services are new, important elements in determining the city's competitiveness in the global economy.

There have been a number of studies attempting to draw a global urban hierarchy according to cities' producer services bases. Leislie (1995) identifies the top international centers of advertising industries: New York, Tokyo, London, Paris, Chicago and Los Angeles. Moss (1987) also draws a map of the worldwide location of US advertising agencies, presenting the dominance of New York, Los Angeles, Chicago, San Francisco and London over the world. The journal *Advertising Age* (27 January 1997) details advertising agencies, advertisers, creative output and people that have made New York the capital of the advertising world over the last century. Moss's maps show the locations of foreign offices of the top 500 US law firms and of offices of the top 13 international accounting firms, highlighting the dominance of London and Paris around the world. Warf (1996) notes the concentration of the largest engineering firms in New York (New Jersey), London, Tokyo and Frankfurt.

Table 3.8 lists cities with large advertising demands. New York and Tokyo form the outstanding first tier in terms of market size, London, Chicago and Paris the second, and some US cities including Los Angeles, Detroit, San Francisco and Minneapolis the third. The prominence of New York and Tokyo in the advertising industry reflects the fact that their markets are about ten times the size of those of Sydney, Düsseldorf, Seoul and Amsterdam at the lower level. The geographical concentration of internationally famous advertising agencies in a handful of global cities is well revealed in Table 3.9. New York, Tokyo, London and, to a lesser degree, Chicago and Paris house most large advertising firms of which worldwide gross income exceeded one billion dollars in 1997. The Omnicom Group, a New York-based ad organization with 30 subsidiaries throughout the world, collected more than half of its billings ($2.2 billion) from the markets outside the US. Tables 3.8 and 3.9 indicate that advertising and image making in the global market are aggregated in a few world cities.

SUMMARY OF COMMAND FUNCTION DATA

All the data suggest the emerging dominance of Tokyo, the relative decline of London and New York, the importance of Paris (rarely mentioned in the existing literature) and the emergence of new centers of global command such as Osaka. Table 3.10 summarizes the data on economic command functions, including corporate headquarters, banks, stock markets and advertising agencies. The dominance of Tokyo is obvious: it is ranked first in the corporations and banks tables, second in the advertising market and third in the stock markets. Together Tokyo, London and New York are the pivotal centers of the world economy, although a strong case could be made for the

Table 3.9 The world's top 20 advertising agencies, 1997

Rank	Ad organization (agency or agency holding company)	Headquarters	Worldwide gross income ($ mil.)
1	Omnicom Group[a]	New York	4,154.3
2	WPP Group	London	3,646.6
3	Interpublic Group of Cos.	New York	3,384.5
4	Dentsu	Tokyo	1,987.8
5	Young & Rubicam	New York	1,497.9
6	True North Communications	Chicago	1,211.5
7	Grey Advertising	New York	1,143.0
8	Havas Advertising	Paris	1,033.1
9	Leo Burnett Co.	Chicago	878.0
10	Hakuhodo	Tokyo	848.0
11	MacManus Group	New York	842.6
12	Saatchi & Saatchi	London	657.0
13	Publicis Communication	Paris	625.0
14	Cordiant Communications Group	London	596.7
15	Carlson Marketing Group	Minneapolis	285.2
16	TMP Worldwide	New York	274.1
17	Asatsu	Tokyo	263.1
18	Tokyu Agency	Tokyo	204.5
19	Daiko Advertising	Tokyo	204.4
20	Abbott Mead Vickers[a]	London	187.3

[a] Omnicom owns 26.8% of Abbott Mead Vickers

Source: *Advertising Age,* 27 April 1998 (http://adage.com/dataplace/archives/dp226.html)

inclusion of Paris, especially given its dramatic increase in the number of major bank headquarters, and perhaps even of Frankfurt which looks on the edge of a breakthrough into the top ranks. There then follow a range of cities which have important ranks in either banks and headquarters (e.g. Munich and Osaka), headquarters and advertising agencies (e.g. Chicago, Detroit and Seoul) or banks (e.g. Beijing, Brussels, Toronto and Milan). The data also allow the identification of emerging world cities; perhaps the best example is Beijing, which is of increasing importance in international banking but has, as yet, to develop a stock exchange or corporate headquarters.

There are other data to be used and more sophisticated analyses to be performed. We have restricted our comments to a small range of data and a simple ranking of information. Much more work is still to be done. Our initial paper which first showed these data (Short *et al.,* 1996) has generated more sophisticated analyses and data sets. Godfrey and Zhou (in press) analyzed the regional subsidiary locations, rather than just the headquarters of multinational corporations, while Beaverstock *et al.* (1998) extend the analysis to producer services, skilled international labor migration and the content analysis of 'business news'.

Our data analysis is at a preliminary stage. However, even at this early stage, we can note that the results, while confirming the dominance of Tokyo, New York and London, have high-

Table 3.10 Global command centers[a]

City	Corporations, 1997	Banks, 1996	Stock markets, 1996[b]	Advertising agencies, 1997
Tokyo	1	1	3	2
New York	2	6	2	1
London	6	4	1	3
Paris	3	2	5	5
Frankfurt	11	3	4	11
Osaka	4	7		
Chicago	7	38		4
Detroit	5	62		7
Munich	8	9		
Amsterdam	9	12		17
Zürich	12	13		
San Francisco	13	18		8
Rome	14	14		
Düsseldorf	15	22		15
Stuttgart	20	21		
Seoul	10	66		16
Milan	36	11		13
Madrid	33	17		
Los Angeles	34	42		6
Beijing		5		
Brussels		8		
Toronto		10		
Boston	40	40		12
Montreal		15		
Charlotte		16		
The Hague	19			
Melbourne		19		
Minneapolis	107	39		9
São Paulo		58		10
Sydney		37		14

[a] Rankings in Tables 3.2, 3.5, 3.7 and 3.8. Cities ranked within the top 20 in the corporation and bank tables, within the top 5 stock markets and within the top 17 advertising agencies are listed in this table

[b] Based on the number of listed companies (domestic and foreign)

lighted several changes over the last few decades, including the increasing dominance of Tokyo and, to a lesser degree, Paris; the relative decline of New York and London; and the emergence of second-tier cities, such as Frankfurt, Osaka and Chicago.

Table 3.11 Central place versus network systems

Central place system	Network system
centrality	nodality
size dependency	size neutrality
tendency towards primacy and subservience	tendency towards flexibility and complementarity
homogeneous goods and services	heterogeneous goods and services
vertical accessibility	horizontal accessibility
mainly one-way flows	two-way flows
transport costs	information costs
perfect competition over space	imperfect competition with price discrimination

Source: Batten (1995: 320)

TELECOMMUNICATIONS INFRASTRUCTURES

The past decade has witnessed a massive injection of new information and communication technologies into the urban economy (Brotchie *et al.*, 1991; Castells, 1989; Castells and Hall, 1994). Intelligent buildings, teleports, fiber optic cables and other leading edge technologies have become part of the emerging infrastructure for the informational city (Batten, 1995; Hepworth, 1990; Langdale, 1991; Moss, 1987, 1988; Warf, 1995). The construction and expansion of these telecommunications infrastructures are critical to the future economic growth of a city and its achievement of world city status. The 'global reach' of WAN (inter-city wide area networks), MAN (metropolitan area networks) and LAN (local area networks) infrastructures creates a new economic and technological basis in the urban competition for markets, resources and employment opportunities (Gillespie and Williams, 1988; Hepworth, 1990). The concentration of the most advanced telecommunications facilities in a handful of world cities is a guarantee of their further prosperity. Alles *et al.* (1994) argue that technological advances in telecommunications in the contemporary world economy are creating a severe technological gap between the largest cities and the rest of the world.

Batten (1995) notes that the global economy is nurturing an innovative class of polycentric urban configurations that he calls network cities. He defines a network city as a city that evolves when two or more previously independent cities, potentially complementary in function, strive to cooperate and achieve significant scope economies aided by fast and reliable corridors of transport and communications infrastructure. Based on the examples of Randstad in The Netherlands and Kansai in Japan, he argues that these creative network cities can have competitive advantages over some of their monocentric rivals named central places cities (Table 3.11).

Although there is a consensus that telecommunications networks play a significant role in the formation and maintenance of world cities, the extent of their influence on repositioning individual cities in the global urban hierarchy has yet to be measured. A number of studies provide maps or diagrams of the global (or regional) network to indicate world cities' hub status: the Pacific Rim fiber-optic links (Warf, 1989), the BITNET and NSFNET in the US (Warf, 1995; Batty, 1991), the international leased network (Langdale, 1989), the Global Digital Highway (Dicken, 1998: 156), the I.P. Sharp Network and the London Stock Exchange Topic Network (Hepworth, 1990), the distribution of teleports (Warf, 1989) and the geography of access to the Internet (Warf, 1995).

There is increasing interurban competition to become the hub of international telecommunications networks (Langdale, 1989). Telecommunications development is well incorporated into cities' marketing strategies: Manchester is aspiring to become Europe's first online city, Amsterdam has labeled itself Informatics City, while Barcelona and Köln have marketed themselves as Telematics City and Communications City, respectively (Hepworth, 1990; Kellermann, 1993). According to Lanvin (1993: 93), 'the strategic new resource in the world economy is information' and 'the key infrastructure is the telecommunications system'.

Even though there is a consensus that telecommunications play a key role in the formation of world cities (Brotchie *et al.*, 1991; Budd and Whimster, 1992; Daniels, 1993; Graham, 1994; Lanvin, 1993; Price and Blair, 1989), the extent of this impact on the international urban system has not been measured. Problems of data availability loom large (Abler, 1991; Kellermann, 1993). Internationally comparable statistics such as those published by the International Telecommunications Union include only national-level data. Moreover, the data do not allow a separation of business-related communication from that of non-business-related communication. Journals catering to the business community, such as *Satellite Communications* and *Telecommunications Policy*, present data about companies rather than cities or even countries. Because telecommunications involve a wide variety of components (telephone calls, faxes, video conferencing, etc.), information flows between cities are difficult to assess. Statistics on individual teleports or computer systems are highly fragmentary. Furthermore, information on private telecommunications networks is for the most part considered confidential (Abler, 1991). This poses a major problem in studying business communications because over half of all international communications now involve internal channels of multinational corporations (Graham, 1994). In many countries private investments in telecommunications equipment today exceed public investments (Lanvin, 1993).

Empirical analyses tend to be restricted to technical aspects of telecommunications such as different types of networks (Hepworth, 1989; Langdale, 1989), or specific networks such as Europe (Cooke and Moulaert, 1992), North America (Hepworth, 1989, 1990), the London Stock Exchange (Hepworth and Ducatel, 1992), Singapore (Corey, 1991) and Japan (Staal, 1994).

The available data are insufficient for ranking cities by the level of telecommunications services they offer. It is possible, however, to note the leading position of the US in private telecommunications services. In 1992 *Satellite Communications* published a private telecommunications networks directory including 460 networks, a growth of 36% in comparison with the previous year. Seventy-five per cent of those were located in the US (Marek, 1992). The *Fortune 500* list includes 22 telecommunications companies in 1998. Nine of these are located in the US, nine in Europe and two in Japan. In part, however, this may reflect the level of privatization of telecommunications in the US rather than the pre-eminence of that country in the global flows of information.

Despite the constant assertion and reiteration of the importance of telecommunications in the global hierarchy, there have been, as yet, very few comparative empirical analyses. The data for a careful comparison are unavailable and the few fragments of available information are insufficient. The global economy hinges upon telecommunications. With more data on business phone calls or electronic mail between cities around the world, which have so far been unavailable, we may get better estimates of interurban information flows. Our inability to measure and compare the flows of information between global command centers is a major problem for research on the global urban hierarchy.

Hubs in the global urban network

According to Smith and Timberlake (1995a), the linkage-based strategy places the overall morphology of the global network of cities and the changing position of cities within this web-like structure at the center of analysis. There are great advantages to conceptualizing cities as nodes in a multilayered network of interactions in that it allows us the opportunity to envision the functioning connectivity of the global urban system.

There are a variety of ways to measure inter-city flows. Smith and Timberlake (1995b) identify four functions of flows (economic, political, cultural and social reproduction), and three forms the flows may take (human, material and information) (Table 3.12). Their 12 conceptual categories of flows suggest the rich variety of inter-city linkages. There is, however, a major problem of data availability.

One of the few excellent data sources of inter-city connections is the information on international air flows. A number of studies have used airline data, particularly the number of daily non-stop flights, to analyze national or regional transport networks: they include the impacts of changes in air service connectivity on employment for large metropolitan areas in the US (Goetz, 1992; Ivy et al., 1995); the airline networks and metropolitan development in the Asia-Pacific region (O'Connor and Scott, 1992); and the crucial role of air transport linkages in the regional development of former Eastern Europe and the

region's integration into the European economy (Ivy, 1995). Data on air links have also been used, with other networks like rail and road, in the analysis of interurban networks to measure the accessibility of individual cities (Bruinsma and Rietveld, 1993; Cattan, 1995a, b). Goetz (1993) and Goetz and Sutton (1997) examine the impacts of deregulation on the US airline industry and on the geography of the US air network.

Globalization is both reflected and embodied in the increase in international accessibility. Because of its capacity to reply rapidly to changes in supply and demand, air traffic provides a good indicator for evaluating the international character of cities (Bruinsma and Rietveld, 1993; Cattan, 1995a; Forsström and Lorentzon, 1991; Hepworth and Ducatel, 1992; Knox and Taylor, 1995). The impact of air transport on cities is similar to that of telecommunications. On one hand, the rapid development of air transport has made distant places more accessible and facilitated the dispersal of economic activities. For

Table 3.12 Conceptualizing inter-city linkages: a typology

Function	Form		
	Human	Material	Information
economic	labor, managers, lawyers, consultants	capital, commodities	business phone calls, faxes, telex messages, technology transfer, advertisements
political	troops, diplomats, social workers	military hardware, foreign aid	treaties, political threats
cultural	exchange students, dance troupes, rock concerts, theater	paintings, sculpture, artefacts	feature films, videos, phono albums (CDs)
social reproduction	families, Red Cross, community organizers	remittances, foreign aid	postcards, night phone calls

Source: Smith and Timberlake (1995b: 86)

example, Boeing 747 jets, widely used today, fly at 640 miles per hour, while the twenty-first century hypersonic aircraft may be able to fly at 4,000 miles per hour (Janelle, 1991). On the other hand, the sophisticated air traffic networks require expensive infrastructure which necessitates their location near big cities. Large international airports are major nodes of economic activities in the global economy.

Keeling (1995) argues for the need to theorize transport's critical role in shaping the evolving world city system. He calls this approach the transport–world city nexus. According to Keeling, airline linkages offer the best illustration of transport's role in the world city system for five reasons:

1   Global airline flows are one of the few indices available of transactional flows or interurban connectivity.
2   Air networks and their associated infrastructure are the most visible manifestation of world city interaction.
3   Great demand still exists for face-to-face relationships, despite the global telecommunications revolution.
4   Air transport is the preferred mode of intercity movement for the transnational capitalist class, migrants, tourists and high-value, low-bulk goods.
5   Airline links are an important component of a city's aspirations to world city status.

There are only a few case studies which explore this transport–world city nexus. Keeling (1995) analyzes global links of world cities on the basis of the number of non-stop and direct flights (the same plane but with one or more stops *en route*) between 266 cities whose metropolitan populations exceeded one million; and Smith and Timberlake (1995a) apply network analysis to data on the number of passengers traveling between pairs of 23 of John Friedmann's 30 world cities (1986).

These studies have increased our knowledge of linkages between cities in the global urban network. However, there is little use of time-series data. We thus have static snapshots rather than pictures of an evolving system. Moreover, most studies on the international air network have used flight frequency data for certain cities. The frequency of airline service is a response to changes in the demand of air passengers. We would argue that the number of air passengers is also an important indicator of the actual volume of flows between cities.

DATA ANALYSIS

We analyzed time-series data for passenger flows between cities around the world. The results are interesting in their own right as well as providing a suggestive picture of an evolving global urban hierarchy.

In order to analyze and map the dominant connections in the global airline network and to show changes to the network over time, data on the number of international air passengers were collected for 1983, 1988 and 1994. The data were provided by the International Civil Aviation Organization (hereafter ICAO), which has published the digest *On-Flight Origin and Destination* since 1977.

This digest of statistics contains information on the revenue traffic carried by airlines on scheduled international services, given in the form of numbers of passengers between pairs of cities (ICAO, 1996a). The revenue traffic data shown for each city-pair in the digest are the combined total of all the reporting airlines. For example, in the 1996 digest for the 12-month period ending 31 December 1994, 127 international scheduled airlines reported a total of 11,192 records of individual city-pairs linked by international scheduled services. No statistics for a city-pair are published unless at least two airlines representing two different states have reported. ICAO has a policy of maintaining confidentiality and excludes data on city-pairs reported by only one airline; 6,024 records were excluded because they were city-pairs reported by only one airline. The rest (5,168 records) were combined to account for the 2,154 city-pairs in 1994. We used the ICAO data for 1983, 1988 and 1994 to analyze the nature of connections between cities around the world. The data can be analyzed in a variety of ways. We have chosen to concentrate on

- the growth of the global air network
- the increasing density of flows in the network
- the emergence of hub cities in the network.

Table 3.13 The growth of international air passenger traffic[a]
(passenger-kilometers flown, billions)

	1984	1994	Average annual growth (%)
Africa	36.0	47.2	2.8
Asia/Pacific	207.8	490.1	9.0
Europe (incl. Russia)	407.9	525.5	2.6
Middle East	41.3	62.0	4.1
North America	513.5	867.2	5.4
Latin America/Caribbean	64.1	105.8	5.1
World	1,270.6	2,097.8	5.1

[a] Based on the registration of airlines of ICAO's member states
Source: ICAO (1996b)

THE GROWTH OF THE GLOBAL AIRLINE NETWORK

The growth of international air flows is indicative of an increase in the cross-border movements of people, which is reflecting and, at the same time, accelerating contemporary globalization. The ICAO data clearly reveal a large increase in international air traffic over the past decade (Table 3.13). Total international traffic generated by the airlines of ICAO's member states was 2,097.8 billion passenger-kilometers in 1994, which shows a huge growth from 1,270.6 billion in 1984. The figures represent a 5.1% average annual growth. The globalized nature of the world economy has generated more business travels across the world, while the growth of international tourism has also played an important role in the rise of international air flows (Hanlon, 1996; Waters, 1995).

Table 3.13 also shows a regional variation in the growth of world passenger traffic. Airlines registered in the Asia/Pacific region, spurred by economic growth and airline liberalization, have experienced the highest rate of growth (Bowen and Leinbach, 1995). In contrast, Africa, the Middle East, and South America and the Caribbean have seen less growth of international air traffic over the years. North America has been the busiest area in terms of international air travel, while Europe, which has developed a sophisticated rail system linking major cities in the region, has decreased its share of international air traffic from 32.1% in 1984 to 25.1% in 1994.

Table 3.14 The growth of the global air network

	1983	1988	1994
Air passengers	103,511,998	172,024,601	217,192,121
Air routes (%)	2,562 (100.0)	3,332 (100.0)	2,819 (100.0)
Air routes with over 100,000 (%)[a]	115 (4.5)	214 (6.4)	299 (10.6)
Air routes with over 300,000 (%)[a]	19 (0.7)	39 (1.2)	61 (2.2)
Cities with heavy connections[b]	72	106	116

[a] Air routes with more than 100,000 or 300,000 passengers per year in each direction
[b] Cities with at least one air route with 100,000 passengers per year in each direction
Source: ICAO (1985, 1990, 1996)

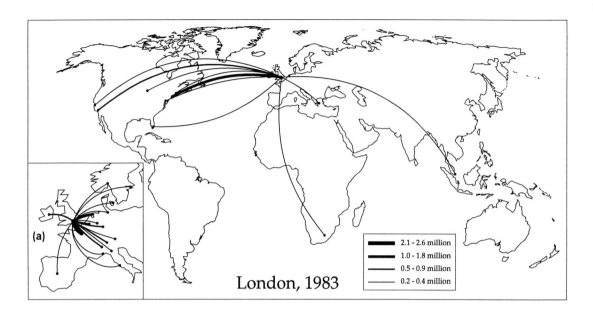

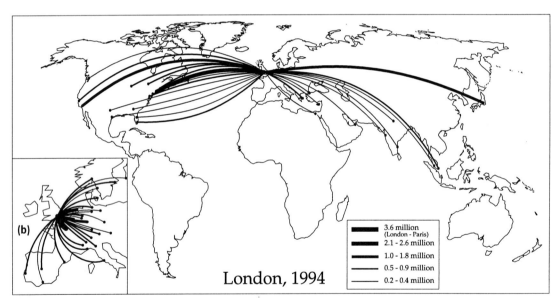

Figure 3.2 Growth in the connection of London.
(*Source*: ICAO, 1983 and 1994)

The growth of international air passenger traffic in the last decade has brought more cities into the expanding web of the global airline network. We can see this in two ways. First, Table 3.14 shows the increases in the volume of passenger flows and the number of total air routes and high-density air routes with more than 100,000 and 300,000 passengers per year in each direction. We selected these two arbitrary levels to concentrate attention on city-pairs having intensive connections. Total air passengers crossing national borders have doubled during the past decade. The share of high-density routes with more than 100,000 passengers in total international air

Table 3.15 International air flows between major city-pairs[a]

City-pairs		Passengers			Absolute growth (1983–94)	Growth rate (%) (1983–94)
		1983	1988	1994		
London	Paris	2,056,205	2,756,401	3,637,159	1,580,954	77
London	New York	1,586,985	2,372,641	2,574,431	987,446	62
Hong Kong	Taipei	905,226	1,337,120	2,462,816	1,557,590	172
Kuala Lumpur	Singapore	1,225,429	1,505,124	2,318,779	1,093,350	89
Seoul	Tokyo	780,838	1,458,375	2,189,387	1,408,549	180
Honolulu	Tokyo	848,791	1,550,932	2,149,091	1,300,300	153
Hong Kong	Tokyo	1,162,564	1,799,214	1,881,383	718,819	62
Amsterdam	London	987,335	1,463,888	1,809,892	822,557	83
Bangkok	Hong Kong	924,933	1,579,541	1,796,119	871,186	94
Jakarta	Singapore	563,591	1,110,507	1,615,182	1,051,591	187
Bangkok	Singapore	624,195	1,066,953	1,490,357	866,162	139
Frankfurt	London	658,541	1,102,855	1,409,946	751,405	114
Hong Kong	Singapore	620,942	811,485	1,382,533	761,591	123
Brussels	London	572,553	852,917	1,207,400	634,847	111
Hong Kong	Manila	582,323	717,909	1,206,597	624,274	107
New York	Paris	773,029	1,136,745	1,150,999	377,970	49
Singapore	Tokyo	314,634	628,313	1,106,410	791,776	252
Taipei	Tokyo	752,354	1,048,066	1,033,441	281,087	37
Los Angeles	Tokyo	660,885	1,130,471	1,023,983	363,098	55
London	Los Angeles	537,352	673,977	1,009,782	472,430	88
London	Tokyo	199,620	490,525	987,847	788,227	395
Hong Kong	Seoul	244,167	455,913	979,681	735,514	301
Frankfurt	New York	815,202	909,248	957,745	142,543	17
Chicago	Toronto	N.A.	931,496	953,525	N.A.	N.A.

[a] City-pairs listed had more than 1 million (rounded number) passengers in both directions in 1994.
Source: ICAO (1985, 1990, 1996)

routes has increased from 4.5% in 1983 through 6.4% in 1988 to 10.6% in 1994; 72 cities around the world were linked to this heavy-loaded airline network in 1983, increasing to 116 by 1994. In the case of air links carrying more than 300,000 passengers per year in each direction, the number of links had tripled over the period. More cities have heavier passenger loads.

Second, we can present the data in map form by concentrating on the pivotal point of the global air network, London. It would be difficult to show all the connections around the world because the information would overload the map. However, by concentrating on the flows from the hub of the network we gain a clear picture of the increasing embrace of dense connections. Figure 3.2(a) and (b) shows the marked growth in the connection of London to cities around the world over the decade. In 1983, London had 27 air routes with more than 100,000 passengers per year in each direction (0.2 million on the maps). By 1994, this had increased to 53 city-pairs. The number of air routes carrying more than 300,000 passengers per year in each direction had risen from five to 22 over the same years. Figure 3.2(b) shows the expansion of London's connection not only to cities in Europe and North America but also to cities in the Middle East, southeast Asia and further northeast Asia. There are increasing global connections, but the spread is uneven. London is clearly linked to cities in Europe, Asia and North America, yet cities in Africa, Central and South America fail to register in the maps.

THE INCREASING DENSITY OF THE GLOBAL AIRLINE NETWORK

There has also been an increasing density of flows within this expanding network. We have seen increases in the density as well as in the geographical coverage of London's air links in Figure 3.2. Table 3.15 shows the world's busiest air routes and their changes over time. The fastest growing air route is London–Tokyo which connects two global cities. Asian city-pairs show substantial growth, accounting for seven of the top ten pairs (and 12 of the top 20), while only four European city-pairs are ranked within the top 20 in 1994. In terms of absolute, as opposed to relative, growth, London–Paris, Hong Kong–Taipei, Seoul–Tokyo, Honolulu–Tokyo, Kuala Lumpur–Singapore, Jakarta–Singapore and London–New York have grown more than others. Alongside some Asian air routes linking neighbor cities within the region, links to/from London, Paris, and New York have gained a large number of passengers over the years. The London–Paris route, the heaviest international air route, absorbed 1.6 million more passengers in the 1983–1994 period, which accounted for 1.4% of the total growth of world passenger flows.

HUB CITIES IN THE GLOBAL AIRLINE NETWORK

How can we explain cities with a large number of air routes and heavy passenger loads? Beyond the size of population, there may be two other possible explanations. Firstly, there are cities with large local demands in terms of origin and destination traffic. Some cities accommodate more internationalized businesses requiring direct air links and frequent services to as many inter-national cities as possible. Amsterdam (Randstad Holland in general), for example, has a prominent international orientation in terms of financial activities, management and distribution functions that generate a large volume of international traffic (Shachar, 1994). Other cities like Honolulu are major destinations of international tourists. Secondly, some cities may function as connecting points, or hubs, which transfer passengers from one air route to the other.

The 'hub–spoke' form of airline route structure is now evident in the international airline network (Button, 1991; Goetz, 1992; Oum et al., 1995). In contrast to a linear network (point-to-point) providing direct services between two specific points, the hub and spoke structure uses one or more cities as regional collection point(s) for passenger flow. Passengers from many different origins are funneled via the spokes into the hub city to connect with flights to their final destinations. This system has led to greater flow efficiency, and has also led to increases in connectivity for cities whose airports are hubs in the configuration. Hub cities have a relatively large number of non-stop destinations available from their airports, and subsequently have generated a tremendous amount of air traffic flow. Hub and spoke networks have been particularly highlighted in the US since deregulation, as major domestic carriers have developed networks concentrating on a few hub cities: examples include US Air's Pittsburgh, Delta Airlines' Atlanta, American Airlines' Dallas–Fort Worth and United Airlines' Chicago (Goetz, 1992; Goetz and Sutton, 1997; Hanlon, 1996; Lee et al., 1994). The international air network has also had hubs benefited by geographical advantages, well-established airport infrastructure or good connections with other transport networks. For example, Singapore and Hong Kong, which have relatively tiny local markets, have achieved hub status in the international air network through connecting air routes between Asia and the rest of the world.

Table 3.16 presents major cities' connectivity based on the numbers of total air routes and air routes with more than 100,000 and 300,000 passengers per year in each direction. The data clearly indicate the dominance of London whose numbers of air routes and combined total passengers dwarf

Table 3.16 City connectivity, 1994

City	Passengers		Air routes connectivity					
	Total	Rank	Total	Rank	Over 100,000	Rank	Over 300,000	Rank
London	40,100,596	1	104	1	53	1	22	1
Paris	22,057,535	2	100	3	35	2	8	3
Tokyo	20,489,342	3	41	13	24	4	13	2
Frankfurt	18,328,218	4	101	2	31	3	3	10
Singapore	15,832,348	5	51	8	21	6	6	5
Hong Kong	15,790,185	6	44	11	16	10	7	4
New York	15,314,455	7	58	6	22	5	6	5
Amsterdam	12,080,055	8	67	4	20	7	2	14
Seoul	10,025,371	9	35	18	13	11	4	7
Bangkok	9,796,801	10	43	12	12	12	4	7
Los Angeles	9,752,336	11	41	13	17	9	4	7
Miami	8,982,601	12	39	15	19	8	1	21
Rome	8,177,195	13	59	5	11	13	3	10
Zürich	7,573,166	14	58	6	9	18	1	21
Copenhagen	6,581,140	15	37	16	10	16	3	10
Brussels	6,467,087	16	45	10	9	18	1	21
Madrid	6,437,612	17	37	16	11	13	2	14
Milan	5,380,135	18	31	19	11	13	2	14
Toronto	5,299,124	19	19	47	10	16	3	10
Kuala Lumpur	5,279,302	20	30	22	5	29	1	21
Top 20 total	249,744,604		1,040		359		96	
(%)	(57.5)		(36.9)		(60.0)		(78.7)	
World total	434,384,242[a]		2,819		598[a]		122[a]	
(%)	(100.0)		(100.0)		(100.0)		(100.0)	

[a] Compared to Table 3.14, world totals are double because combined data of incoming and outgoing direction were used for each city's passengers and air routes
Source: ICAO (1996)

those of all other cities. Air passengers from/to London account for 9.2% of total passengers traveling around the world.

Table 3.16 also shows the top 20 cities ranked by the total number of air passengers from/to a city's airport or airports. The top 20 cities' share of world total air routes is 18.4%, whereas the cities account for 57.5% of total air passengers and 60.0% and 78.7% of heavy routes with more than 100,000 and 300,000, respectively. These top cities have a large number of heavily loaded air routes relative to other cities in the global air network. Figure 3.3 shows the world's busiest air links having more than 1 million passengers in both directions in 1994. The greatest number of passengers are mainly on medium- or short-haul flights; 18 out of the top 24 city-pairs are mainly regional connections. The long-haul routes that have the largest volume of air traffic are London–New York, New York–Paris, Los

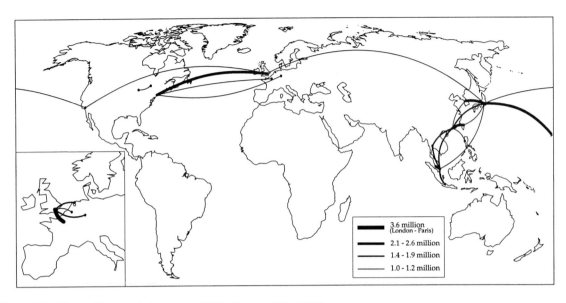

Figure 3.3 The world's busiest air routes in 1994. (*Source:* ICAO, 1994)

Angeles–Tokyo, London–Los Angeles, London–Tokyo and Frankfurt–New York. These cities, frequently named world cities in world cities research, not only have massive local demands but also transfer intercontinental passengers to smaller cities around the world.

Compared to Asian cities, European cities tend to have more diverse but less loaded air routes. The European airlines have traditionally had the highest share of international passengers; almost three-quarters of their clients cross national borders. American domestic hub cities with mammoth airports like Chicago, Atlanta and Dallas do not stand out in the international airline network. Fewer than 10% of passengers arriving at and departing from Chicago O'Hare, the world's busiest airport, are international, yet more than 80% of passengers in London Heathrow, Frankfurt and Paris Charles de Gaulle are heading for or coming from foreign cities (Hanlon, 1996: 135).

Below London, the indisputable hub of the global airline network, two categories of cities can be identified. First, there are dominant world centers in the network. The data in Tables 3.17 and 3.18 suggest that Paris and Frankfurt followed by Tokyo and New York play an important role in connecting cities around the world, showing high centrality and connectivity relative to other cities. Each of them handles more than 15 million international air passengers each year, and has greater than 20 high-density links delivering more than 100,000 passengers. Their high-density links reach at least four continents around the world. This finding contrasts with previous studies. Keeling (1995), for example, asserts New York, London and Tokyo as dominant world cities in the global airline network. However, even using his own data reveals inconsistencies: his tables on direct airline services and weekly non-stop flights to major world cities rank Tokyo in the tenth and fourth position, respectively. In those tables Paris is placed second and third and Frankfurt third and fifth, but these two cities are eventually positioned in a lower tier than Tokyo. Regardless of his data, it seems, Keeling repeats the oft-cited notion that New York, London and Tokyo sit at the top of the global urban hierarchy. Given the high centrality and connectivity of Paris and Frankfurt in the international air network, we argue that these two cities should also be considered as major world centers in the global urban hierarchy.

Second, below those world centers, there are more regionally based centers such as Singapore, Amsterdam, Miami, Los Angeles and Hong Kong.

Table 3.17 Air links by city and region[a]

City	Europe	North America	Central and South America	Middle East	Asia	Oceania and Pacific	Africa	Russia and Former Eastern Europe	Total
London	29	14		3	6		1		53
Paris	23	7		2	3				35
Frankfurt	13	9		1	5			3	31
Tokyo	4	5			11	4			24
New York	13	2	4		2	1			22
Singapore	2	1			13	5			21
Amsterdam	18	1			1				20
Miami	3	1	15						19
Los Angeles	3	3	3		8				17
Hong Kong	2	3			10	1			16
Seoul		3			9	1			13
Bangkok	4				7	1			12
Madrid	8	2	1						11
Milan	10	1							11
Rome	9	1			1				11
Copenhagen	9	1							10
Toronto	2	8							10

[a] Air routes with more than 100,000 passengers per year in each direction
Source: ICAO (1996)

These cities' connections are greatly centered on their own regions, while London, Paris, Frankfurt, Tokyo and New York are more globally connected. The distinction between global and regional centers is confirmed by the geographical coverage of the cities' high-density air routes. The regional articulation of these third-level centers' connections is shown in Figure 3.4(a)–(e). Figure 3.4(a) clearly demonstrates that Singapore is the regional hub of southeast Asia and Oceania. Amsterdam's connections are mainly oriented to European cities, and Miami evidently shows its inclination to Central and Latin America. Grosfoguel (1995) and Nijman (1996) point to Miami's exceptional characteristics in the sense that the city combines a subordinate and peripheral position in the national urban system with a prominent international position as the 'Capital of Latin America'. Los Angeles is connected with most of the Pacific Rim cities, while Hong Kong plays an important role in linking Asian cities to European and American cities. These regional centers serve as hubs in their regional networks and as connecting points between major world cities and small cities in their regions.

A city's position in the global airline network can rise and fall over time. We have noticed in Table 3.15 a dramatic increase of air traffic in Asian city-pairs and some links between major world cities. Table 3.18 shows changes in major cities' centrality in the network which are indicative of an evolving global urban hierarchy. A city's centrality in the international air network has been measured by the number of the city's air routes carrying more than 100,000 passengers per year in each direction. The number of high-density

Table 3.18 Air routes: the evolving global urban hierarchy, 1983–1994[a]

City	1983 Air routes	Rank	1988 Air routes	Rank	1994 Air routes	Rank	Absolute growth (1983–94)	Growth rate (%) (1983–94)
London	27	1	43	1	53	1	26	96
Paris	14	3	20	3	35	2	21	150
Frankfurt	7	7	20	3	31	3	24	343
Tokyo	13	4	19	5	24	4	11	85
New York	15	2	24	2	22	5	7	47
Singapore	9	5	12	6	21	6	12	133
Amsterdam	6	10	8	13	20	7	14	233
Miami	5	11	12	6	19	8	14	280
Los Angeles	5	11	10	9	17	9	12	240
Hong Kong	7	7	12	6	16	10	9	129
Seoul	3	21	8	13	13	11	10	333
Bangkok	3	21	4	27	12	12	9	300
Milan	4	16	9	10	11	13	7	175
Rome	5	11	7	15	11	13	6	120
Madrid	4	16	7	15	11	13	7	175
Copenhagen	7	7	9	10	10	16	3	43
Toronto	0	–	7	15	10	16	10	43[b]
Zürich	8	6	9	10	9	18	1	13
Brussels	3	21	5	21	9	18	6	200
Athens	5	11	6	19	8	20	3	60
Osaka	5	11	6	19	8	20	3	60
Sydney	2	29	7	15	7	22	5	250
Stockholm	4	16	5	21	7	22	3	75
Chicago	1	39	5	21	7	22	6	600
San Francisco	3	21	4	27	7	22	4	133
Mexico City	3	21	5	21	6	26	3	100
Taipei	4	16	4	27	6	26	2	50
Vancouver	3	21	4	27	6	26	3	100
Barcelona	0	–	3	35	6	26	6	100[b]
Buenos Aires	0	–	2	47	6	26	6	200[b]

[a] Air routes with more than 100,000 passengers per year in each direction; rank of 114 cities in 1994, 102 in 1988 and 67 in 1983
[b] 1988–94; data were 0 for 1983
Source: ICAO (1985, 1990, 1996)

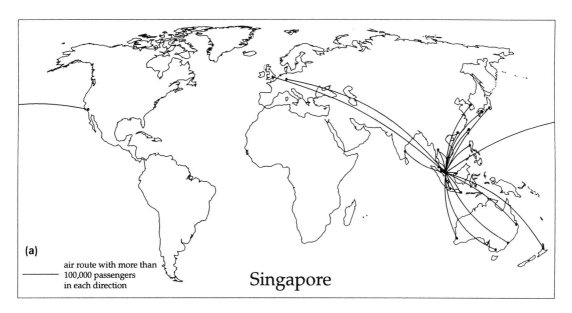

(a)

air route with more than
100,000 passengers
in each direction

Singapore

Figure 3.4 Regional centers: (a) Singapore; (b) Amsterdam; (c) Miami; (d) Los Angeles; (e) Hong Kong. (*Source*: ICAO, 1994)

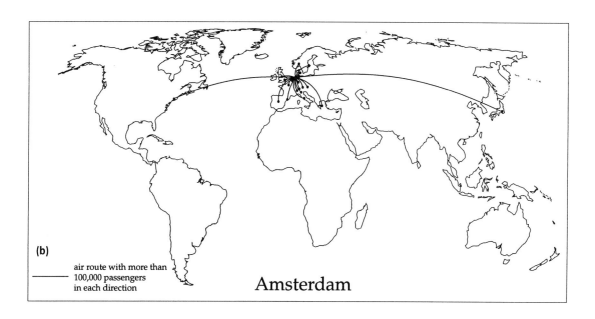

(b)

air route with more than
100,000 passengers
in each direction

Amsterdam

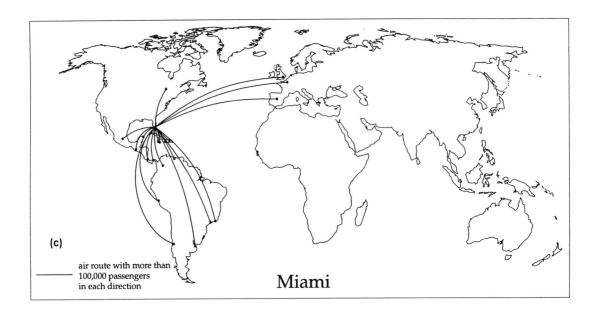

(c)

air route with more than
100,000 passengers
in each direction

Miami

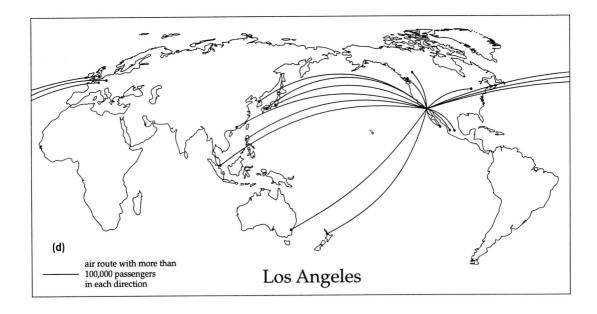

(d)

air route with more than
100,000 passengers
in each direction

Los Angeles

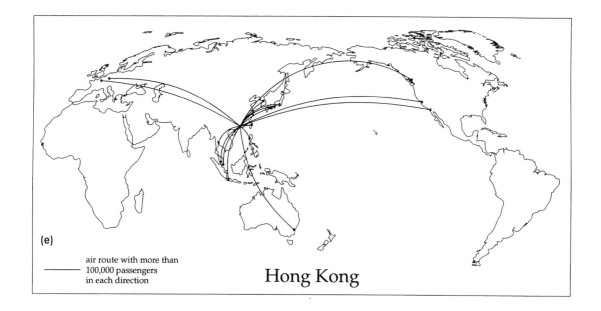

(e)

air route with more than
100,000 passengers
in each direction

Hong Kong

air routes may be more meaningful for examining the city's functional nodality and connectivity than that of all routes or of passengers. London has been the prominent center in the global air network and has gained the greatest number of heavy links over the years. Unlike Tokyo and New York showing moderate growths, Paris and Frankfurt have largely expanded their geographical reaches and intensified their connections to other cities. Frankfurt's stretch to the former Eastern European cities (such as Moscow, Warsaw and Prague) has contributed to the remarkable growth of that city's centrality. Miami, Amsterdam and Los Angeles among regional centers have subsequently risen to serve as hubs in their respective regions over the past decade.

Below these global, world and regional centers in the hierarchy, the fastest growth rates have been experienced by Seoul, Bangkok and Toronto, while Brussels, Sydney, Chicago, Barcelona and Buenos Aires have also shown rapid growths. In contrast, Copenhagen, Zürich and Stockholm have substantially lost their regional bases in Northern Europe.

In conclusion, we have shown that London has been the prime hub of the global airline network. The next tier consists of Paris, Frankfurt, Tokyo and New York. Frankfurt and Paris have largely outpaced New York and Tokyo over the past decade in terms of the number of air routes with heavy passenger loads. Regional centers such as Singapore, Amsterdam, Miami, Los Angeles and Hong Kong are ranked in the third tier. These cities have functioned as regional hubs linking neighbor cities to major world centers.

Compared to the global urban hierarchies suggested by many world cities researches concentrating on the city's commanding functions in the global economy, our findings are of interest in three ways. First, we have shown that Paris and Frankfurt have played an important role in the global urban system, serving as hubs of the international air network. These two cities, however, have been downgraded in many other rankings (Friedmann, 1995; Keeling, 1995). Second, our maps for the five regional centers clearly demonstrate their different geographical orientations. We have shown that the geographical coverage of those centers is more regional than global. Third, the changing picture of the global air network suggests that major cities, such as global, world and regional centers, have gained more passengers and air routes than most other cities around the world over recent years. As the nodes of the global urban network, these hub centers are facilitating the transnational interactions between cities and being in turn favored by the increased interactions in the process of globalization.

Much work remains. The data can be analyzed with more sophisticated techniques that can measure changing accessibility and connectivity for the whole system and individual cities. We have introduced a useful data source revealing the linkage between cities around the world and a meaningful method to illustrate a rapidly changing global urban hierarchy.

4
World Cities

The world occupies its capital cities as a renter rather than an owner, and even when it buys it is usually prepared to sell and move on.
(Rieff, 1991: 19)

The hub points in the global network are world cities. The term 'world cities' was first coined by Patrick Geddes in 1915; he defined them as places where the world's business was done. As a general definition it is a good one. Today, there is a vast array of finer-grained definitions of world cities. Friedmann (1986) defines world cities as:

- basing points in the spatial organization and articulation of production and markets
- major sites for the concentration and accumulation of international capital
- centers of corporate headquarters, international finance, global transport and communications, and high level business services
- points of destination for both domestic and international migrants.

Sassen (1991, 1994) identifies them in four ways:

- highly concentrated command points in the organization of the world economy
- key locations for finance and specialized service firms
- sites of production, including the production of innovations, in leading industries
- markets for the products and innovations produced.

Shachar (1994) adds world cities' rich provision of physical and social infrastructure, and Mollenkopf and Castells (1991) point to social polarization as an important characteristic of world cities.

There has been a consensus in the literature of world cities on the distinctiveness of London, New York and Tokyo in terms of their supremacy in the contemporary world economy. Although the economic conceptualizations of world cities are very fruitful for understanding their role in a global economy, Grosfoguel (1995) argues that there are other global logics which also explain the emergence and multiple functions of world cities in the world system. They include the geopolitical/military and the ideological/symbolic logic in which cities like Washington DC and Miami play a significant role in their respective arenas.

Many comparative analyses of world cities have stressed the commonalties of London, New York and Tokyo in their economic dominance, spatial restructuring and social polarization. However, Markusen and Gwiasda (1994) argue that New York is distinctly dissimilar from London and Tokyo, both because it operates in a much more decentralized and multipolar national urban system, and because long-term deindustrialization weakens its transactional portfolio, a trait completely different from Tokyo's. Knox (1995) points to Tokyo's particular position in the global urban hierarchy, as the city's status as a world city is based very largely on its economic dimensions, with little of the cultural diversity or dynamism of London and New York, and nothing comparable to their legacies of political and military hegemony. Unlike the two other cities, London has a long history of a colonial, imperial city (King, 1990a).

While London, New York and Tokyo have

widely been accepted as world cities at the top-most level, there has been little consensus in the literature on which cities should be included in the next tier. Friedmann (1995) identifies Miami, Los Angeles, Frankfurt, Amsterdam and Singapore as second-order centers; while Yeung (1996) and Lo and Yeung (1997) include all the big cities in Asia: Seoul, Taipei, Hong Kong, Manila, Bangkok, Kuala Lumpur, Singapore and Jakarta, as well as Tokyo. There is no consensus in the literature.

Nijman (1996) points to a discrepancy between a city's ranking in the national and international urban hierarchy. He claims that Miami combines a subordinate and peripheral position in the US urban system with a prominent and central role in the Caribbean region. Los Angeles, San Francisco, Washington DC, Boston and Chicago have been named world cities in many studies (Abu-Lughod, 1995; Lyons and Salmon, 1995; Markusen and Gwiasda, 1994).

In a special issue on globalization and the US city system (*Urban Geography*, Vol. 17, No. 1), Warf and Erickson (1996) note that cities other than London, New York and Tokyo have received remarkably little scrutiny in the context of globalization. Ward (1995) also addresses the failure of much contemporary work on world cities to look seriously and systematically at large cities in less developed economies.

Cities which are competing for world city status, are called 'wannabe world cities' (Short, 1996), 'world cities on the edge' (Todd, 1995) and 'the world's next great international cities' (Rutheiser, 1996). Friedmann (1995) notes an ongoing competitive struggle for position in the global network of capitalist cities and the inherent instability of this system. One way of looking at the fierce competition for world city status is to examine cities which have applied for hosting global spectacles such as the Olympic Games. The list of host and candidate cities for the Summer Olympic Games is presented in Table 6.3 of Chapter 6. Cities like Seoul, Melbourne, Nagoya, Barcelona, Amsterdam, Atlanta, Toronto, Manchester, Sydney and Beijing have hosted or applied to host the Games during the past decade. Cities that applied for the 2004 Summer Olympics include Athens, Buenos Aires, Cape Town, Rome and Stockholm. Hosting the Olympic Games is a way to boost international recognition and achieve world city status. The 1996 Centennial Summer Olympics, for example, provided a great opportunity for rethinking Atlanta, boosting the city's economy, and advertising the image of a healthy, vital and integrated city through the unprecedented international publicity and media attention. The Atlanta Committee for the Olympic Games claimed that 'the Olympics would establish Atlanta as one of the top cities in the world; right up there with the Parises, and the Tokyos, and the New Yorks' (Rutheiser, 1996: 285).

World cities are central nodes in the global network of economic flows and particular socio-cultural places. In this chapter we will briefly mention three aspects of world city status: economic dominance, socio-cultural infrastructure and social polarization. We will then look at how these aspects play out in the globalization of Seoul.

Command centers

World cities are control, command and management centers that orchestrate global manufacturing production, financial transactions, producer services and telecommunications networks. They house the most crucial institutions of economic globalization, such as transnational corporate headquarters, stock markets, advertising agencies and teleports. Dieleman and Hamnett (1994) note that because world-city status is a guarantee of enhanced levels of prosperity in the contemporary world economy, the status is something to be coveted and defended.

Table 4.1 clearly demonstrates that the areas receiving most attention in world cities research include the disproportionate concentration on financial and producer services and the associated spatial restructuring. The table also shows a difference in the attention given to aspects of the three cities. Tokyo and London, compared to New York, have been identified by their strong primacy in their national economies. The concentration of transnational corporate headquarters in Tokyo has been highlighted. With the worldwide consumption of American cultural products, New York's cultural centrality has been emphasized. In

Table 4.1 Major foci in world cities research

City	Studies	Corporate headquarters	Finances	Producer services	Telecommunications	Social infrastructure	Social polarization	Spatial restructuring	Cultural hegemony	Primacy in national economy
London	Budd and Whimster, eds (1992)		●				●	●		●
	Frost and Spence (1993)			●						●
	Graham and Spence (1995)			●						●
	Graham and Spence (1997)			●				●		
	King (1990a)	●	●	●			●	●		
	Meyer (1991b)		●							
	Pryke (1991, 1994b)		●					●		
	Thrift (1994b)		●			●				
New York	Beauregard (1991)							●		
	Bergesen (1991)								●	
	Longcore and Rees (1996)		●	●	●			●		
	Lyons and Salmon (1995)		●							
	Markusen and Gwiasda (1994)		●	●						
	Mollenkopf and Castells, eds (1991)			●			●			
	Moss (1991)			●	●					
	Noyelle, ed. (1989)		●			●				
	Sassen-Koob (1986), Sassen (1990)						●			
	Scanlon (1989)		●	●	●			●		
	Shefter, ed. (1993)		●						●	●
	Warf (1991)		●	●						
Tokyo	Fujita (1991)	●	●		●			●		●
	Machimura (1992)	●						●		
	Masai (1989)				●			●		●
	Rimmer (1986)	●								●
London & New York	Fainstein et al. (1992)		●	●			●	●		
	Zukin (1992)		●					●		
All three	Sassen (1991)	●	●	●	●	●	●			●

aggregate terms, whereas most world cities researchers mention Tokyo's increasing strength with the growth of the Japanese economy, Tokyo itself has received rather less attention in terms of the relationship between globalization and urban changes.

Socio-cultural infrastructure

The concentration of top-level financial institutions and producer services in a few world cities cannot be fully explained by economic factors alone. The success, vitality and buoyancy of world cities cannot be reduced to narrow economic factors of agglomeration and accessibility. Pryke and Lee (1995) note the importance of the 'socially constructed cultural geography' of financial production in understanding competitive relations between and within financial centers. Similarly, Budd (1995) argues that the comparative advantage of world cities largely depends on the maintenance of 'global credentials'.

Amin and Thrift (1992, 1994) call for a deeper understanding of the importance of social make-up and various traditions in the construction and maintenance of world city status. Their argument is not that economic factors are unimportant, but that social and cultural factors are crucial to economic success. They call those factors the 'institutional thickness' of world cities. In some places, such as the City of London, this favorable institutional thickness has persisted over a long period of time. Amin and Thrift identify three important elements. First, centers provide the face-to-face contact needed to generate and disseminate discourses and collective beliefs. Second, centers are needed to enable social and cultural interaction, that is, to act as places of sociability, of gathering information, establishing coalitions, monitoring and maintaining trust, and developing rules of behavior. Third, centers are needed to develop, test and track innovations, to provide a critical mass of knowledgeable people and structures and socio-institutional networks, in order to identify new gaps in the market, new uses for and definitions of technology, and rapid responses to changes in demand patterns. These centers of geographical agglomeration are 'centers of rep-

resentation, interaction and innovation'. Thrift (1994b) terms the characteristic culture of global financial centers 'information, expertise and contacts'. Knight (1995) also argues the significance of strengthening all aspects of the city's cultural base, especially knowledge culture, for city development in an open and increasingly knowledge-based global society.

Social polarization: winners and losers

World cities contain some of the richest but also some of the poorest members of society. While the rise of financial and producer services has produced a pool of high-paid jobs, the number of low-income households has also expanded in many cities. In their world-city hypothesis, Friedmann and Wolff (1982) and Friedmann (1986) have identified social polarization as a major impact of the increasing capital intensity of production and large-scale immigration of foreign workers in world cities. Saskia Sassen's (1991) comparative study of New York, London and Tokyo highlights the paradoxical relationship between the ascendance of finance and producer services and the growth of an informal economy in these cities. She argues that the shifts from manufacturing to services since the mid-1970s have caused a widening gap in wealth between classes, because services produce a larger share of low-wage jobs than manufacturing does, and several major service industries like producer services also produce a larger share of jobs in the highest-paid occupations (Sassen-Koob, 1986; Sassen, 1990, 1991, 1994, 1995).

The notion of world cities containing the extremes of wealth and poverty has also been illustrated in the 'dual city' of Mollenkopf and Castells (1991) and the 'divided cities' of Fainstein et al. (1992). The depiction of New York by Mollenkopf and Castells (1991: 3) can be best summed up in a sentence: 'Wall Street may make New York one of the nerve centers of the global capitalist system, but this dominant position has a dark side in the ghettos and barrios where a growing population of poor people lives.' Morita and

Sassen (1994) note the role and importance of increasing illegal immigration in creating a divided city, Tokyo, that previously has been known as a relatively homogeneous city.

Parallel to these findings on the existence of the poorest of the poor within London, New York and Tokyo, some empirical studies have revealed that global cities continue to attract disproportionate flows of skilled international migrants. Beaverstock and Smith (1996) point to a large magnitude of cross-border movements of financial specialists within the London-based investment banking industry, while Findlay *et al.* (1996) point out the skilled international labor migration in Hong Kong expatriate communities. Since banking professionals and managers are crucial to the control of the globally integrated financial system, their presence is both a pre-requisite and consequence of being a world city.

There are some disagreements on the thesis of social polarization in world cities. Abu-Lughod (1995) raises fundamental questions on polarization in world cities: are inequalities greater in today's world cities than they used to be, and more so in world cities than in non-world cities? And if they are, by what mechanisms are such inequalities linked to increased integration with the global economy and its restructuring? She argues a need for more controlled empirical investigation on social polarization in world cities. Hamnett (1994, 1996) criticizes Sassen's arguments regarding the nature and causes of polarization in terms of the over-generalization of American cases to all global cities around the world. Under the title 'Why Sassen is wrong' (1996), he argues that the existence of strong welfare programs in many European cities, in his case Randstad Holland, has meant that global economic pressures have not led to the absolute growth of a large low-wage service economy. Instead, it has resulted in growing unemployment and a large and growing state-dependent population. Hamnett (1994) prefers 'professionalization' to polarization. Murie and Musterd (1996) also note the wider functioning of the welfare state in Dutch cities. A well-funded welfare program is an important reason why Dutch cities, and more broadly European cities, have not exhibited such marked social segregation as American cities.

Becoming a world city: the globalization of Seoul

Most research on world cities has concentrated on cities in Europe and North America and the emphasis has been on economic globalization. In this section we examine urban changes accompanying economic, cultural and political globalization in Seoul, a 'wannabe world city' in Asia. In particular, we assess the influences of globalization on changes to Seoul's economy, culture and politics, and Seoul's local initiatives to rewrite global trends.

SEOUL'S GLOBALIZED ECONOMY

Over the past decade Seoul's connection to the world has markedly expanded and intensified. Global economic forces, such as multinational firms, the World Trade Organization (WTO) and the Organization for Economic Cooperation and Development (OECD), have put tremendous pressure on Seoul to assure free capital flows into the Korean economy. In addition, Seoul-based firms and banks have started investing abroad in the 1990s.

As shown in Table 4.2, the inflows of foreign direct investment in Seoul increased about nine times between 1982 and 1996. Most of these investments have concentrated on the service sector, including hotels, stock markets, financial firms and trading companies.

Many well-known international hotels have set up their foreign affiliates in Seoul, such as The Seoul Renaissance Hotel, The Swiss Grand Hotel Seoul and The Ritz-Carlton Seoul. The penetration of foreign capital into the Korean retail and wholesale markets can be observed in the proliferation of American fast-food chains and convenience stores in Seoul. McDonald's is now one of the favorite sites for schoolchildren in Seoul, and since a 7-Eleven set up in 1989, 385 convenience stores, financed by mainly joint ventures of American or Japanese companies and local entrepreneurs, have opened up in Seoul (The Seoul Metropolitan Government, 1995b).

Joining the OECD (December 1996) has been a very important factor in accelerating Seoul's

Table 4.2 Foreign direct investment in Seoul, 1982–1996

| | FDI flows ($ million)[a] | | | | National share, |
	1982	1987	1992	1996	1996 (%)
Manufacturing	133.3 (22.7)	291.7 (19.0)	512.8 (17.5)	666.7 (13.4)	8.0
Services	451.8 (76.9)	1,245.8 (80.9)	2,421.4 (82.5)	4,310.3 (86.6)	69.0
Hotels	212.1	872.9	1,186.8	1,245.8	53.9
Finance	146.6	216.8	605.4	1,122.1	81.3
Commerce and trade	0.4	11.5	178.6	558.1	86.8
Retail and wholesale	0.5	16.6	61.0	283.5	50.4
Insurance	3.0	4.4	181.9	237.4	99.8
Total	587.4 (100.0)	1,539.2 (100.0)	2,934.9 (100.0)	4,977.7 (100.0)	33.9

[a] Outstanding as of the end of each year
Source: International Economic Policy Bureau, Ministry of Finance and Economy, Korea (1996)

market and financial liberalization. Membership obligations based on the OECD codes have forced the Korean government to open up to foreign traders and investors (Ley and Poret, 1997). Indeed, Korea's experiment with heavy-handed state capitalism officially came to an end in 1996, when the government abandoned the 35-year-long Five-Year Economic Plans and switched to a neoliberal, market-oriented economic model.

The securities market in Seoul has shown the most progress in terms of financial liberalization. Thirty-five foreign security companies had set up branches in Seoul by 1997. Most of them have opened up relatively recently. Foreigners were allowed to invest in Korean stocks only as recently as January 1992; initially they could directly own up to 23% of listed stocks, but in May 1997 the limit on foreign stock ownership (ceiling) was expanded to 50%. Stock ownership by foreigners increased sixfold during 1990–96 and continues to increase apace in the IMF era. These banks and security and insurance firms have significantly contributed to the rapid increase of inward FDI flows in Seoul.

Although Seoul's FDI outflows were initiated as early as 1968 when Korea Nambang Development invested in Indonesia for lumbering, foreign direct investment had not been common to Korean firms until the mid-1980s. Large surpluses in the trade balance in the 1985–88 period,

nationwide labor disputes in 1987 combined with wage increases, and the government's deregulation on capital exports since 1993 have all acted as 'push factors' in the relocation of Korean firms abroad. Korea's total outward FDI had increased over 2,600%, rising from $112.8 million in 1985 to $3,058.9 million in 1995. Developing countries in Asia, particularly China and southeast Asia, are favored sites for Korean firms' overseas investment. Low wages, relatively easy penetration into local markets and rich natural resources in these countries have been major locational factors for Korean firms (Singh and Siregar, 1995).

A few Seoul-based multinationals, including Samsung, Hyundai, Daewoo and LG, have led this explosive growth of Korea's FDI outflows. Their business strategies, so-called 'global management', have played an important role in shaping the Korean people's understanding of the notion that 'Globalization is a must' (*Far Eastern Economic Review*, 1995). Their 'beyond-Asia' investments in Africa, East European countries and Russia prove the companies' truly 'global' management and expanded global reach. Indeed, this surge of overseas investment over the past few years has resulted in the 'liquidity crisis' that was a crucial factor in the recent Korean financial troubles. Many Korean banks and firms have 'unwisely' channeled their short-term foreign loans into these long-term investments abroad. Their inability to

Table 4.3 Distribution of the overseas offices of Seoul-based banks and financial firms, 1996

	Banks			Financial firms[a]			Total
	Branches	Agencies[b]	Representative offices	Branches	Agencies[b]	Representative offices	
Hong Kong	6	20	14	–	23	21	84
New York	12	9	10	2	11	19	63
London	8	6	7	–	15	20	56
Tokyo	12	–	5	6	1	18	42
Los Angeles	6	17	2	1	–	1	27
Singapore	7	3	6	1	2	7	26
Jakarta	–	8	2	–	5	2	17
Frankfurt	2	6	3	–	–	–	11
Shanghai	3	–	2	–	–	6	11
Top nine cities	56	69	51	10	57	94	337
World total	90	96	95	11	83	113	488

[a] Includes security, insurance, investment and trust, and lease firms
[b] Includes agencies and individual branches' local offices
Source: Ministry of Finance and Economy, Korea

meet the interest payments due on these short-term loans finally resulted in the IMF rescue package in late 1997 and early 1998.

Along with the deregulation of capital outflows and the acceleration of overseas direct investment, Korean banks and securities, insurance, lease and trust and investment firms have rushed into the global financial market (Table 4.3). In June 1997, 97 branches, 103 agencies and 87 representative offices from Seoul's 17 banks were operating in 56 cities throughout the world. More than half of these foreign banks, specifically offices in China, southeast Asia and post-Socialist states, were set up in the 1990s. Forty-seven out of 72 foreign offices of Seoul's insurance companies have been established since 1990, yet all 113 foreign offices of security firms and 11 of the trust and investment companies began their overseas businesses in the 1990s.

In April 1996, the Korean government lifted all outflow limits, which had stood at one billion won (about $1.1 million) for corporations and 500 million won for individuals (*Far Eastern Economic Review*, 1996b). The government implemented this measure to coincide with the latest increase of the foreign stock-ownership ceiling. It wanted to offset additional funds from increased foreign investment with an appropriate amount of domestic outflows in order to maintain a sound capital account balance. In the same context, the government is also relaxing restrictions on Korean companies issuing overseas securities. All of these measures were not even thinkable in previous decades. The Korean financial market is expected to completely open within a year based on the 'advice' of the IMF, which has rescued the continuing devaluation of the Korean won with the global record bailout in late 1997.

The globalization of Seoul's economy has not been restricted to the flows of capital. Cross-border movements of people have also increased. Seoul had functioned as *the* destination point for domestic migration up until 1990. While the magnitude of domestic immigration has decreased, the influx of (il)legal foreign workers into Seoul has largely increased over the past few years.

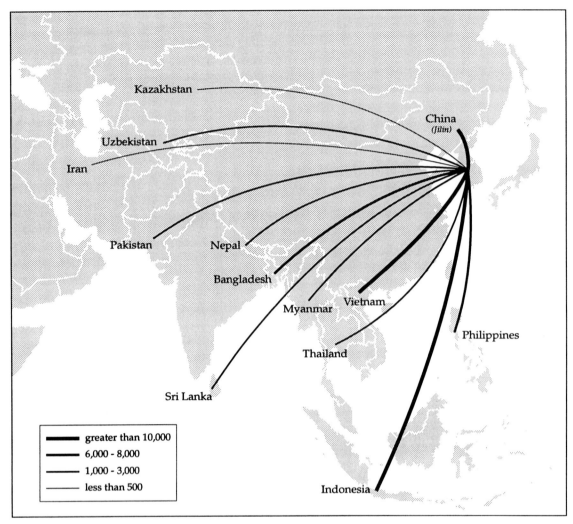

Figure 4.1 Labor migration to Seoul: number of foreign workers for the Industrial Training Programs, November 1993 to February 1996. Jilin in China is the province where most Korean Chinese live and the major origin of Chinese workers in Seoul. (*Source*: Korean Federation of Small Business, 1996, 'The industrial training program for foreigners', 12 November)

Although foreigners account for less than 2% of Seoul's total population – a very low level relative to other world cities around the world – their presence is certainly strengthening Seoul's image as an international city.

The appearance of foreign workers in Seoul began after the 1988 Seoul Olympics when the Korean economy enjoyed fast growth and when Seoul received enormous international publicity through numerous Olympic events. The normalization of political relationships with China in the late 1980s has also spurred the influx of Korean-

Chinese workers in Seoul. Before the arrival of these workers, foreigners accounted for less than 0.01% of Korea's total population, and their jobs were generally limited to soldiers (American overseas army), language instructors, engineering technicians and entertainers.

Acknowledging both the economic contribution of foreign workers to the Korean economy and the mounting potential for conflict between local people and foreigners, the Korean government instituted the Industrial Training Program in 1993 which aimed at legalizing the status of foreign

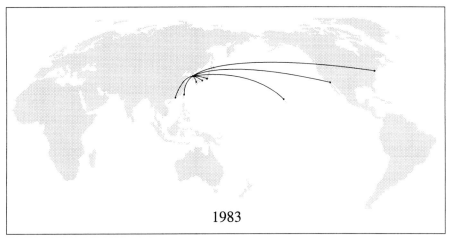

1983

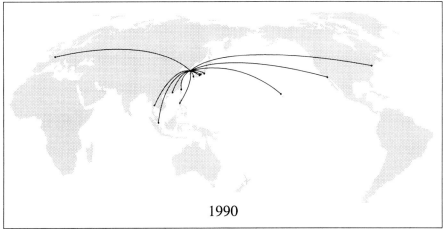

1990

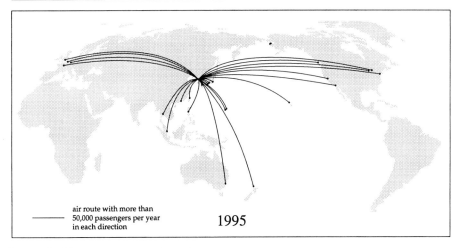

air route with more than
50,000 passengers per year
in each direction

1995

Figure 4.2 The growth of Seoul's air linkage, 1983–1995. (Source: ICAO, 1985, 1991, 1997)

workers but, at the same time, specifying the jobs they could work in and the length of time they could stay in Korea. The Program allows recruited foreign workers to work in some unskilled routine jobs. Through this training program, 60,809 workers from 13 countries throughout developing Asia, particularly China, Indonesia, Vietnam, the Philippines and Bangladesh, had arrived in Korea by the end of 1996 (Figure 4.1). Small manufacturing firms have been the primary beneficiaries of the Program. Indeed, the use of cheap foreign labor contributes to prolonging the profitable existence of those structurally weak industries in the Korean economy.

The government's estimate of 110,751 illegal foreigners in 1996, based on the visa records of overstay, is an underestimate. A more accurate figure is 200,000. The great majority came from China, accounting for 40% of the total, followed by the Philippines, Bangladesh, Thailand and Pakistan. They are mainly employed in the informal economy, ranging from antiquated factories to small restaurants and illegal prostitute clubs. Intermediaries, generically called the 'immigration industry' and involving a diverse group of recruiters, lawyers, travel agents and job brokers, have prospered by encouraging and facilitating illegal migration between Seoul and many Asian developing countries.

In terms of global transport connections, Seoul has not been a prominent center in the international air network. Unlike Hong Kong or Singapore, Seoul is not geographically well situated as a connecting point bridging Asia and other regions. The neighboring presence of Tokyo has also placed Seoul in a weak position to attract international airline services. Despite all of these constraints, Seoul has thickened and widened its air linkages to cities across the globe, mirroring the rapid globalization of the Korean economy over the past decade. As we saw in Table 3.18 in Chapter 3, Seoul's position in the global air network has been sharply elevated over the past decade. The number of passengers departing Seoul has increased by 320.6% between 1983 (1,298,132) and 1994 (5,459,494) (ICAO, 1983, 1994). The city's rank has moved up from 21st to 11th over the corresponding period in terms of the number of heavy-loaded air routes with more

than 100,000 passengers in each direction per year.

Alongside intensifying links to the world, Seoul has also expanded its geographical reach. Figure 4.2 clearly shows that Seoul has broadened its regional coverage of air connections. In 1983, the heavy links of Seoul were all concentrated on cities in northeast Asia and the US. Over the past decade, many routes between Seoul and European and North American cities have gained a large amount of air traffic. Seoul's connection has also stretched to cities in southeast Asia and Australia. The number of air routes carrying more than 50,000 passengers per year in each direction from/to Seoul has increased from eight in 1983 through 13 in 1990 to 24 in 1995. The growth of Seoul's air linkage has reflected and embodied the fast globalization of the Korean economy, which has spurred international travel for business, tourism and academic activities.

Seoul has also been connected to the rest of the world via satellite and undersea fiber-optic cables. As the future of global connectedness clearly lies with optic cables, the Korean government has channeled massive investment into the construction of the optical cable networked to the world (Jeong and King, 1997; Larson, 1995). In 1990, Korea's submarine optical cable was connected to the network linking Hong Kong, Japan and Korea (the APCN). Alongside the construction of the optic links, the liberalization and privatization of telecom services is a very important determinant in the growth of Seoul's telecommunications network. The liberalization has led to the introduction of highly sophisticated business communication services, such as WorldSource Services, into Seoul (AT&T, 1995). This telecom service company has been targeting for multinational companies, which need high-volume data and voice transfer on a global scale. Korea Telecom has deployed WorldSource Services to provide multinational firms operating in Seoul with high-speed data transmission and a single point of contact for ordering, installation, billing, maintenance and network management.

This expansion of fiber-optic networks has spurred the widespread use of the Internet among the younger generation in Korea. Based on Network Wizards' Internet domain survey (1997), the number of hosts under domain 'KR' – most Korean hosts having this domain name – was estimated at 132,370 in July 1997. The Korea Network Information Center (KRNIC) has also counted the number of Korean Internet hosts over the years. The size of the Internet in Korea has increased from 7,650 hosts in 1993 to 131,005 hosts in 1997.

The Seoul economy has been globalized at a fast pace. The inward FDI flows in Seoul have multiplied in the past decade. Many foreign banks and financial firms currently operate in Seoul. The globalization of the Seoul economy is expected to be notably accelerated by the IMF bailout package that followed the recent financial crisis. The IMF has requested the rapid and complete opening up of the Korean financial market.

The explosive growth of Seoul-based corporations' overseas investment and banks' foreign financing is also a clear indication that Seoul is trying to find its own way to secure its competitiveness in world markets. In conclusion, the globalization of the Seoul economy has been reinforced both by global pressures to open up the market and by local initiatives seeking international competitiveness.

SEOUL'S GLOBALIZED CULTURE

Over the past decade Seoul has undergone dramatic changes in its culture as well as its economy. Seoul is home to over 10 million individuals yet the city has a homogeneous character. Single ethnicity, common language and geographical isolation of the entire country have been important factors in maintaining a coherent cultural tradition between Seoulers. In a global era, however, this homogeneous, somewhat insulated, national identity, 'Koreanness', has been challenged by external influences. As the Seoul economy opened up to the world, Seoulers began to consume more goods produced abroad. They now come across foreigners on the street more often than ever before. Many Hollywood films are being released in Seoul and many American cultural products, ranging from women's peroxide dyes to Burger King franchises to the latest NFL results, are readily available in Seoul.

Seoul has used and continues to use global sport events to reinforce its claims as a center of international importance and to enhance its international profile. Due to the global reach of TV media, specific international events staged at particular venues are concurrently consumed across the globe. The Olympic Games and the soccer World Cup, the most widely watched media events in today's world, have drawn huge television audiences and countless web site visitors around the world as well as a large number of stadium spectators.

The Seoul 1988 Summer Olympics have generally been considered as a statement of the political and economic accomplishment and aspiration of Korea, and as a statement, by contrast with North Korea, of the superiority of a capitalist model of development over its communist counterpart (Larson and Park, 1993). During the years leading up to the Olympics, Seoul worked hard to craft a distinctive, appealing image for itself. These images and identities of Seoul signified throughout the Olympic period were very pervasive, and they have mostly obscured Seoul's war-torn image, held by many people around the world since the Korean war.

The Seoul Olympics also played a significant role in enabling Seoulers to taste and appreciate other cultures. Numerous programs for the preparation of the Games prompted Seoulers to learn more about other cultures and languages (SLOOC, 1989). The Olympics drew hundreds of thousands of foreign visitors from around the world. National teams from 160 countries participated in the Games, 30 of those countries having had no official foreign relations with Korea until then (Kim et al., 1989). The physical presence of these foreigners in Seoul, although it lasted for a very limited time, inspired Seoulers to diversify their homogeneous culture and their US- and Japan-oriented foreign relations.

The Seoul government has maintained its commitment to the widening and deepening of Seoul's cultural diversity in the post-Olympic decade. When Seoul won the right to co-stage the first FIFA World Cup in Asia, the city had already drafted a number of grand projects in the hope of achieving the status of 'a true world city'. The national and city governments have proposed the 'greater' and 'faster' internationalization of the

Figure 4.3 Cultural globalization in Seoul. (*Photograph by John Rennie Short*)

policy, aimed at encouraging more domestic production, had been accused by Hollywood film makers of contradicting the free-trade policy. After long negotiations between the US and Korea over the distribution rights of Hollywood films, the Korean film market was finally opened up in the fall of 1988 to major Hollywood film companies such as United International Pictures (UIP), Twentieth Century Fox and Warner Brothers. These companies have subsequently expanded their market share to distribute more than a third of Hollywood films imported by 1995.

The invasion of American film companies has resulted in a phenomenal increase in the release of Hollywood films in Seoul, while local film industries have dramatically declined (see Figure 4.4). The Koreans spent $7.7 million on importing foreign films in 1987 and the spending had increased to $67.9 million by 1995. In 1995, 80% of revenues in the Korean movie market were directed to foreign films, particularly Hollywood films.

Many American TV shows, such as *Home Improvement*, *The Cosby Show*, *ER* and *Beverly Hills 90210*, have been shown in Seoul. A broad range of documentaries on historical events, exotic culture and African wilderness, mainly supplied by the Discovery channel in the US, cover a significant block of prime time TV. Alongside these imported programs, many sitcoms and dramas in Seoul have imitated popular American shows in terms of contents, characters and even settings. The sitcom *Three Guys, Three Girls* on MBC, receiving the highest rating over the past few years, is very similar to *Friends* on NBC in the US. It is not difficult to find the 'Korean versions' of *60 minutes*, *General Hospital* and *NYPD Blue* in Seoul.

In Seoul, many professionals have been educated abroad, especially in the US. In the period 1982–97, a total of 15,350 doctors of Korean nationality had been trained abroad, and 61.2% of them had received their degrees from universities in the US. The size of Korean student enrollments in US universities has increased from 6,150 students in 1980/81 to 36,231 in 1995/96 (Institute of International Education, 1980/81, 1995/96). In 1996/97, Korea was the leading

Seoul economy and culture (The Ministry of Information and Communications, Korea, 1996).

The globalization of Seoul can also be observed in popular culture (Figure 4.3). Hollywood films have been dominant in the global cinema market, and Seoul has been no exception to this global trend. The release of Hollywood films in Seoul has outnumbered that of Korean movies since 1989 (Figure 4.4). In 1995, Seoul imported 202 films from Hollywood, 50 from Hong Kong, and 107 from the rest of the world, while only 64 movies were produced by local film makers. In competition with Hollywood films, the relative decline in the popularity of Hong Kong films has been commonly observed throughout Asia (Dissanayake, 1996). Hollywood films have also outpaced Korean ones in attracting audiences since the 1970s. Hollywood films have accounted for 7–8 of the top 10 grossing films in Seoul each year.

Up until 1988, the Korean cinema market had been highly regulated. In order to import foreign films, local film companies were required to make a certain number of their own films. This quota

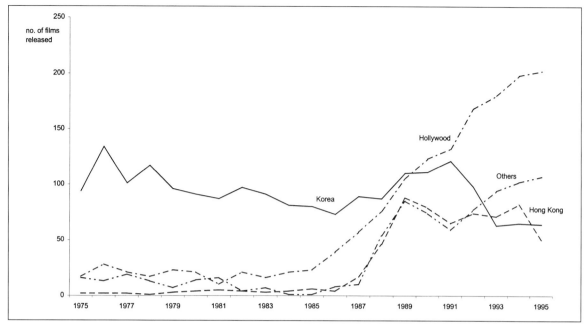

Figure 4.4 Dominance of Hollywood films in Seoul. (*Source*: Ministry of Culture and Sports, Korea)

home country for Intensive English Program (IEP) students across the US, as 10,226 Korean students (23.4% of the IEP total) enrolled in various American language institutes to learn English (*op cit.*, 1996/97). Many multinational firms and producer services companies operating in Seoul prefer professionals with American educational backgrounds, who can speak English and presumably have an 'international sense'.

Although these Korean expert cultures do not simply reflect American values, they have played a significant role in the promotion of 'American lifestyle' in Seoul. Their behaviors, ideals, tastes and habits, which are by and large detached from ordinary Koreans' lives and partly influenced by their experiences in the US, have been considerably reflected in a broad array of their professional work, such as national economic policies, architecture, urban planning, college lectures, legal practices, business management, shopping patterns and fashion.

It has been debated whether the increasing interconnectedness of economic, cultural and political activities across the globe would undermine local people's territory-bounded identity (Mlinar, 1992; Stevenson, 1997). There has been a general

agreement that the current accelerated phase of cultural flows through the active movements of people, commodities, technologies, ideas, capital and images around the world may not refer to the triumph and sovereign domination of any single metanarrative (Appadurai, 1996). Stevenson (1997) argues that national cultures remain more powerful constructions than many postcolonial and postmodern critics seem aware.

Over the last decade Seoulers have experienced multiculturalism through their diversified economic, cultural and political connections with other countries around the world. The rise of foreign travel and the growing presence of foreign workers in association with the hosting of the Olympics have played a significant role in the diversification of Seoul's cultural exchanges. Soaring demands on the geographical and cultural knowledge of foreign matters have led to the proliferation of foreign language schools and travel agencies for overseas travel.

Although Seoul has not been a major destination of international tourism, tourist inflows in Korea have always outnumbered its outflows in the past three decades. The Korean government had strongly regulated its citizens' cross-border

Table 4.4 Popular Hollywood films in Seoul and the US

(a) The top ten grossing films in Seoul and the US, 1993–1995

	Rank	*1993*	*1994*	*1995*
Seoul	1	Cliffhanger	The Lion King	Die Hard with a Vengeance
	2	Jurassic Park	True Lies	Forrest Gump
	3	Su-Pyun-Je[a]	Speed	Leon (The Professional)[b]
	4	The Bodyguard	Two Cops[a]	Braveheart
	5	Aladdin	Schindler's List	Stargate
	6	Home Alone 2: Lost in New York	Forrest Gump	French Kiss
	7	The Fugitive	The Piano[b]	Seven
	8	The Piano[b]	Mrs Doubtfire	Dr. Bong[a]
	9	Demolition Man	Color of Night	Interview with the Vampire
	10	The Last of the Mohicans	Demolition Man	Killing Wife[a]
US	1	Jurassic Park	Forrest Gump	Toy Story
	2	Mrs Doubtfire	The Lion King	Batman Forever
	3	The Fugitive	True Lies	Apollo 13
	4	The Firm	The Santa Clause	Pocahontas
	5	Sleepless in Seattle	The Flintstones	Ace Ventura: When Nature Calls
	6	Indecent Proposal	Dumb and Dumber	Goldeneye
	7	In the Line of Fire	Clear and Present Danger	Jumanji
	8	The Pelican Brief	Speed	Casper
	9	Schindler's List	The Mask	Seven
	10	Cliffhanger	Pulp Fiction	Die Hard with a Vengeance

(b) Popular Hollywood films in Seoul compared to the US, 1993–1995

	Rank in Seoul	*Rank in the US*
Cliffhanger	1	10
Die Hard	1	10
Leon (The Professional)[b]	3	73
Braveheart	4	17
French Kiss	6	43
The Piano[b]	8 and 7	38
Color of Night	9	70
Demolition Man	9 and 10	18

[a] Korean films
[b] Produced in other than the US or Korea
Sources: Ministry of Culture and Sports, Korea; Willis (1994–1996), *Screen World*

66 GLOBALIZATION AND THE CITY

movements and their spending abroad for a long time. In the late 1980s, however, as many Korean firms began to search for new business opportunities abroad, there were enormous demands for travel to foreign countries. Increased concerns about other countries, mainly stemming from the Olympic experiences, also generated increased demand for foreign travel. There was an explosion of overseas travel in the 1990s. In 1995, Korea's tourism outflows eventually exceeded that of its inflows. The number of travelers abroad quadrupled between 1988 and 1995. The tourism balance of payments has become negative since the early 1990s.

While the rise of foreign travel has clearly helped to generate a certain level of openness to other cultures among Seoulers, there is some evidence suggestive of the resilient presence of Koreanness. We draw on two examples which indicate the maintenance and, to some extent, revival of Korean national identity, style and taste among Seoulers in the age of accelerated and intensified globalization: the Koreanization of Hollywood films and the active assertion of nationalist sentiments in various events celebrating Seoul's 600th anniversary as the capital of Korea.

Although the hegemonic presence of the US entertainment industry in Seoul is unquestionable, it does not mean that Seoulers have exactly the same taste as Americans. Table 4.4(a) and (b) shows the top 10 grossing films in Seoul and the US in the past few years. The table indicates that most films ranked within the top 10 in Seoul have been produced in Hollywood. Many films, however, that were very successful in the American market attracted much smaller audiences in Seoul than would have been expected. In contrast, some films, such as *Cliffhanger*, *Die Hard with a Vengeance*, *Color of Night*, *Braveheart*, *French Kiss* and *Demolition Man*, recorded unexpected hits in Seoul, making up for their disappointments in American box offices. In general, action films starring internationally famous actors and lightweight dramas with some sorts of humor, love and tears have strongly appealed to viewers in Seoul.

The fact that Seoul's viewers have very different preferences among Hollywood films from Americans is suggestive of the imperfection of the intrusiveness and pervasiveness of American commodified culture in Seoul. Even if Seoulers love to see Hollywood films instead of not-good-enough local films, their tastes still reflect Korean tastes.

The second example of the contemporary resurgence of nationalism in Seoul is a series of events celebrating the 600th birthday of Seoul as the capital of Korea (16 September to 29 November 1994). Seoul had been considerably restructured by its 'Sixth Centennial Celebration Project' in terms of the cultivation of various cultural spheres, a re-emphasis on the historical continuity from old Seoul to the contemporary city, and the inauguration of numerous community-oriented activities to preserve historic heritage and traditional culture (The Seoul Metropolitan Government, 1995c). The Project focused on tracing the historical roots of Seoulers and constructing local consciousness in the face of global forces (see Figure 4.5). The Seoul Metropolitan Government (1995c) identified four future images of Seoul, including 'an historical city, a humane city, a cultural city and a world city'. These images suggest that the city is pursuing local as well as global dreams. The Sixth Centennial Celebration Project has clearly established that both the 'Seoul Renaissance' and the 'Seoul Globalization' would be strongly emphasized in the reimagining process of the city for the next decade.

In conclusion, Seoulers have been culturally globalized over the past decade. The Seoul Olympics stimulated Seoulers' interests in other cultures. American cultural products have increasingly been introduced into Seoul and have shaped many young Seoulers' cultural tastes. Increasing foreign travel, a globalizing local economy and the government's progressive promotion of internationalization discourse have all contributed to the recent rise of cultural flows between Seoul and the wider world. Nationalist sentiments, however, have been strengthened equally by the very factors that contributed to the formation of global consciousness. The maintenance and cultivation of 'Koreanness,' has an increased rather than a diminished significance in contemporary Seoul.

Figure 4.5 Reconstruction of tradition in central Seoul. (*Photograph by John Rennie Short*)

SEOUL'S POLITICAL GLOBALIZATION

Like many other cities throughout the world, Seoul has implemented a series of growth-oriented economic policies in an attempt to improve its position in world markets. The agenda of 'international competitiveness' has shifted the contour of Seoul's political discourse, both within official policy circles and within public accounts.

The government, either central or urban, has revealed somewhat contradictory natures in the internationalizing process of Seoul. On the one hand, it has been involved in the promotion of Seoul's economic and cultural globalization, as we have seen in the two preceding sections. The Seoul government, on the other hand, has been transformed and weakened by the processes of globalization, as global forces have generated many unexpected, uncontrollable situations, such as the recent financial crisis.

We begin with an investigation of the recent rise of Seoul's cross-border cooperation, with particular attention to the development of the Sister-City Project and the increasing host of international conventions. We then examine the government's responses to the current financial collapse.

The Seoul Metropolitan Government has made tremendous efforts to create Seoul's image of an internationally oriented city (The Seoul Metropolitan Government, 1995d). The government has recognized the considerable contribution of Seoul's international visibility to the advancement of its competitiveness in the world market. The Sister-City Project is clearly part of these efforts. The Project began in 1968 when Seoul and Taipei entered into a twinning agreement in order to share their similar political situations. Seoul, however, had not actively participated in the twinning of sister cities by the late 1980s. Massive cultural exchanges with other countries during the Seoul Olympic period have immensely stimulated Seoul's sister-city movement in the 1990s. Presently, Seoul has twinning relationships with 16 cities in 15 countries around the world and more than half of these twinnings began after the Olympics. Sister cityhood with (former) Socialist cities, such as Moscow, Beijing, Ulaanbaatar, Hanoi and Warsaw, has been highlighted over the past few years, as many Korean firms have sought market and investment expansion in those cities.

As part of its Sixth Centennial Celebration, Seoul hosted the 'Seoul Sister Cities Folk Music and Dance Festival' in October 1994 (The Seoul Metropolitan Government, 1995c). In 1995, Seoul carried out a total of 52 official projects with 11 selected coupled cities, including the joint use of technical expertise and the exchange of observation of urban administrative institutions and affiliates. The Seoul–Tokyo coupling, the most active pair of Seoul's multiple twinnings, for example, had 13 programs for strengthening their sister relationship in 1995. Exchange visits between the two cities involve organizations ranging from environmental groups through professionals in urban sewerage to high school students.

Hosting various international conventions is another important way to promote a city's international status and expand its political connection to other cities around the world. When delegates from all over the world convene at a single site to discuss a specific problem of worldwide importance, the site might be considered a world city (Janelle, 1991). In that regard, many emerging, wannabe world cities in developing countries are eager to host international conferences to demonstrate their importance in world discussions and their role in resolving international problems.

Seoul hosted 67 international meetings in 1996; the city ranked 22nd in the world in this regard. Although Seoul has held far fewer meetings than some top-ranked cities, such as Paris, Vienna, London, Brussels and Geneva, its position has noticeably moved up in the hierarchy of conventional cities over the past few years. As the host city of the 2000 Asia–Europe Meeting (ASEM) and the 2002 World Cup, Seoul is constructing two mammoth international convention centers in South Seoul, Kangnam (Korea National Tourism Organization, 1997). Fully acknowledging the economic, political and cultural benefits of international conferences for the host city, Seoul even enacted a law named 'The Law for the Promotion of International Convention Industry' at the end of 1996.

We can also examine the political consequences of Seoul's globalization by noting the effects of the financial crisis which began in the fourth quarter of 1997. The Korean government was accused of political corruption and inexperience with global economic trends (*The Economist*, 1997g). The IMF bailout package to rescue the Korean economy from the crisis discredited the Korean government's economic policies and practices. The IMF strongly requested the government to follow 'free market principles', which negate much of Seoul's own way of relating politics with corporations and banks. The government, influ-

enced by economic, cultural and political globalization, has already been in transition. The presence of the IMF is expected to accelerate and intensify this transition.

In numerous analyses of the Asian financial crisis in newspapers, economic journals and business magazines, two confrontational interpretations are identifiable. One argues that much of the financial distress is of Asia's own making, and nowhere is this clearer than in Korea (*The Economist*, 1997g). The other interpretation, popular with a much smaller audience, points to the huge inflows and sudden withdrawals of foreign speculative investments in those Asian countries (Passell, 1997; Sachs, 1998). Certainly the two arguments are not exclusive, but complementary in explaining how the Asian tigers have declined so quickly. But each provides very different solutions out of the financial troubles.

For years, governments in Asia have used the banks as tools of state industrial policies, ordering them to make political loans to uncreditworthy companies. These problematic relations among governments, corporations and banks in Asia have led to the overbuilding of industrial capacity and made the banks insolvent and vulnerable to competition with foreign banks. The crisis has prompted calls for a thorough restructuring of their banking systems, a process that agencies like the IMF are certainly willing to assist.

The second interpretation finds the major source of the crisis in international factors, instead of each country's domestic problems. It points to a huge speculative infusion of foreign cash into the banks and stock markets of these Asian countries in the past few years. All of these countries were forced by the WTO to lift restrictions on free capital flows in the early 1990s. The large influx of foreign capital has encouraged excessive borrowing and led to speculative booms in the real estate market throughout Asia. As the currencies in Asia appreciated in recent years and exports leveled off, foreign investors withdrew their money at the same time, highlighting the lack of foreign reserves in these countries.

In the case of Korea, the world's 11th-largest economy and the biggest victim yet of Asia's financial crisis, depreciating currency, falling stock prices, failing corporations and a contentious

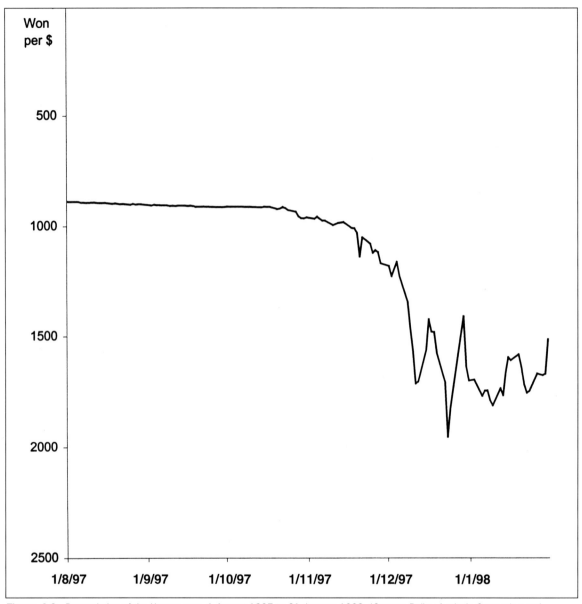

Figure 4.6 Depreciation of the Korean won, 1 August 1997 to 31 January 1998. (*Source*: Policy Analysis Computing and Information Facility in Commerce, University of British Columbia, 1998, 'Pacific exchange service', http://pacific.commerce.ubc. ca/xr/data.html)

presidential election all occurred during the fall and winter of 1997. The widespread phrases, reflecting Seoulers' global dreams, such as 'international competitiveness' and 'world city status', became ridiculed by both the local population and the wider world. The won (the Korean currency) has dropped 89.2% against the dollar from 895.0 in August 1997 to 1,693.7 in January 1998 (Figure 4.6).

Korea asked for IMF aid on 21 November 1997 to uphold its nearly depleted foreign currency reserves (less than 20 billion dollars) and pay back short-term foreign debts (more than 150 billion dollars). Korea and the IMF reached a 56

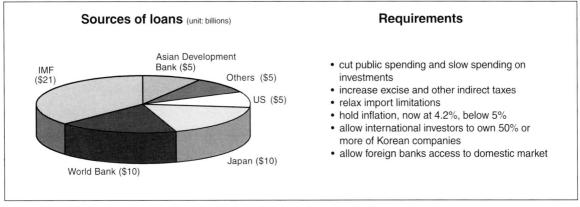

Sources of loans (unit: billions)

IMF ($21)

Asian Development Bank ($5)

Others ($5)

US ($5)

Japan ($10)

World Bank ($10)

Requirements

- cut public spending and slow spending on investments
- increase excise and other indirect taxes
- relax import limitations
- hold inflation, now at 4.2%, below 5%
- allow international investors to own 50% or more of Korean companies
- allow foreign banks access to domestic market

Figure 4.7 The largest global bailout and its requirements. (*Source*: Pollack, 1997b: D6)

billion dollar loan package, the largest international economic rescue ever, on 4 December of the same year. Through the IMF bailout, the Korean government has hoped to stabilize its troubled financial system, quiet the fears of investors and depositors, and re-establish the confidence of the Korean people and foreign investors.

The IMF rescue package, however, is expected to involve a substantial transformation in Seoul's established business and financial practices (Figure 4.7). Strong government intervention, which has, until recently, been considered the core of Korea's success story, would certainly be the first target of the IMF's austerity measures. In exchange for the bailout, Korea has been forced to agree on a set of terms that would clearly cut its economic growth in half for the next few years, raise unemployment sharply, cut government spending and allow more foreign financial institutions access to its financial market.

Seoul's responses to the crisis since the winter of 1997 have distinctly reflected the dwindling capability of a state in the control of its domestic financial system. It is, notes Passell (1997), 'really hard to keep your house in order when foreign currency is easy to obtain at bargain rates', and even harder when it is withdrawn at one time. Indeed, the hardest part is yet to come, as the IMF will stay in Seoul in the next few years and provide a wide range of advice from government spending to layoffs. The IMF has tried to ensure that the Korean government follows the 'market principle'

and 'free international trade' thesis. This story seems to confirm the extreme globalization thesis: the decline of the state in the age of globalization. The fact should not be dismissed, however, that it is the government that is trying hard to reorganize the failing economy, and Seoulers expect their government's pivotal role in the recovery of the sickly economy more than ever before. In one sense, the state is getting bigger, carrying the Seoul economy in a globalizing world.

In conclusion, three aspects have been noted. First, we examined Seoul's globalized economy through foreign investment inbound and outbound, financial capital inflows and outflows, foreign labor migration from developing countries in Asia, and transportation and telecommunications networks connected to the wider world. As the Seoul economy has become more exposed to the global economy, a dramatic increase in the flows of goods, capital, people and information between Seoul and the rest of the world has been observed.

Second, Seoul's transnationalized culture was examined in terms of the host of global spectacles, such as the Olympic Games and the World Cup, the penetration of American cultural products, particularly Hollywood films, TV shows and professional sports, and the rise of foreign travel. The recent revival of national consciousness, 'Koreanness', is a local reaction to the overwhelming trend of Americanization.

Third, transformations in Seoul's urban politics on contact with a globalizing world were examined with respect to pairing arrangements with other cities and the staging of international conferences. The city government's diminishing role in managing the recent financial crisis, a negative consequence of globalization on Seoul, was also discussed.

This was one case study. In order to fully understand the urban effects of globalization and the globalization effects of cities, a broader and deeper range of case studies are needed. The unraveling of the global–urban connections is likely to be a major focus of interest for some time.

PART THREE

Cultural Globalization and the City

5

Cultural Globalization

There's an important distinction between intelligibility and identity ... As English grows it will diverge increasingly and there'll be more varieties of it.

(David Crystal quoted in Cornwell, 1997: 42)

A globalized culture has arisen through 'a continuous flow of ideas, information, commitment, values and tastes across the world, mediated through mobile individuals, symbolic tokens and electronic simulations' (Waters, 1995: 126). The world has been culturally globalized by the phenomenal development of electronic media and the growth of immigration and tourism over the past few decades. Progress in economic globalization has also contributed to the acceleration of cultural globalization. Appadurai (1990, 1996) proposes five dimensions of global cultural flow:

- ethnoscapes (the movement of tourists, immigrants, refugees and guestworkers)
- mediascapes (the worldwide distribution of information through newspapers, magazines, TV programs and films)
- technoscapes (the distribution of technologies)
- finanscapes (global capital flows)
- ideoscapes (the distribution of political ideas and values, such as the master-narrative of the Enlightenment).

The recent growth of these flows, argues Appadurai, has given rise to the formation of a globalized culture, while, at the same time, the disjuncture between different dimensions of the flows has allowed the multiplicity of global culture across the globe.

Global connectedness has also been reinforced by international organizations, religions, international competitions and spectacles, and global entertainment industries. The ascendance of 'global babble' and global discourse is both a reflection and an embodiment of cultural globalization. Rising global consciousness can easily be read off various movements of human rights, feminism and environment which pursue their goals and actions on a global scale. Companies are also advertising their global reach, service and market. The construction of globality is a dominant cultural discourse in the contemporary world.

The main debate

The debates on cultural globalization have polarized into whether the recent surge of cultural flows and global consciousness have increased or decreased sameness between places around the world. The tension between cultural homogenization and cultural heterogenization is the most controversial issue in the interpretation of increasing interactions across the globe (Appadurai, 1990, 1996). In what follows, we briefly review the two main accounts of a culturally globalizing world.

The cultural homogenization account points to the formation of a global consumer culture in the era of late capitalism. Some have identified cultural homogenization as the process of 'cultural imperialism' or 'Americanization' throughout the world (Barnet and Cavanagh, 1996). The facts that people across the globe are watching CNN and MTV, that McDonald's franchises are opening around the world, and that many Hollywood films dominate the world film market, are taken as indisputable evidence of the Americanization of the world. CNN news in English is available in many countries outside the US. McDonald's operates more than 23,000 restaurants in 110 countries and is ceaselessly entering new cities and countries (McDonald's Corporation, 1998). The gross earnings of the academy-awarded movie *Titanic* have exceeded one billion dollars in box offices outside the US (Internet Movie Database, 17 May 1998). The global reach of American cultural products provides large and complex repertoires of images, narratives and values to users and viewers around the world, in which the world of commodities and the world of news and politics are profoundly mixed (Appadurai, 1990: 299). In the age of global media, argue Morley and Robins (1995) and Wark (1994), we experience a 'wired identity' through the transnational trafficking of (dis)orienting images, signs and identities.

Alongside this popular culture, the development of 'global metropolitanism' by the practices of transnational producer service classes, such as international lawyers, architects, copywriters and financial analysts, has been erected (Featherstone, 1990; Knox, 1996; Sklair, 1996, 1998). World cities are more like each other than non-world cities in their respective countries. The work of signature architects such as Frank Gehry, I. M. Pei, Ceaser Pelli, James Stirling and Richard Rogers can be found in prestige metropolitan sites around the world. Frank Gehry, for example, has designed a wide range of buildings across the globe, including the American Center at Paris, the Fishdance Restaurant at Kobe in Japan, the Walt Disney Concert Hall at Los Angeles and, more recently, the Guggenheim Museum in Bilbao, Spain (Kevin Matthews and Artifice, 1998). Buildings designed by these internationally famous architects have made cities around the world look similar and postmodern.

An alternative thesis argues, however, that while particular television programs, sport spectacles, network news, advertisements and films may rapidly encircle the globe, this does not mean that the responses of those viewing and listening within a variety of cultural contexts and practices will be uniform (van Elteren, 1996; Featherstone, 1993, 1995; King, 1997). In a special issue on global culture in the journal *Theory, Culture and Society* (1990, Vol. 7, no. 2–3), many writers argue that the variances in responses to globalization processes clearly suggest that there is little prospect of a unified global culture, rather there are global cultures in the plural. Specific cultural backgrounds are not just empty containers for the receipt of global messages, they are critical to how messages are received and consumed.

The heterogenization argument maintains that people in the contemporary world have become increasingly familiar with the presence of different cultures, rather than sucked into a single cultural orbit. Although the recent explosion of cultural flows and global consciousness is very suggestive of the formation of a globalized culture, the strong presence of indigenous, traditional cultural traits throughout the world has made it difficult to claim that the world has become culturally unified.

Appadurai (1996) writes of the subversive micronarratives (indigenization) that reinterpret and reinvent global messages in manifold local settings. Through examples of the Ainu in Japan and the Congolese in the Congo Republic, Friedman (1990) examines the way in which groups in various national contexts handle consumer commodities in a variety of strategies to reconstitute their cultural identities. The gradual appearance of Puerto Rican vernacular architecture in the Lower East Side of New York City is one example of the distinctive expressions that ethnic communities do not follow strictly the metanarrative of an emergent world city (Sciorra, 1996). Sciorra argues that, given economic, political and social marginality, the Latino and African–American working poor struggle to change the existing conditions in which they live, by creating spaces of their own design that serve as locations of resistance to a system of inequity and domination.

Hannerz (1997: 127) proposes the concept of creolization, suggesting the existence of various scenarios for cultural transformation in the periphery facing compelling cultural forces from the core. He notes that 'When the peripheral culture absorbs the influx of meanings and symbolic forms from the center and transforms them to make them in some considerable degree their own, they may at the same time so increase the cultural affinities between the center and the periphery that the passage of more cultural imports is facilitated.' Hannerz studied a town in Nigeria where oil money brought TV to the townspeople. Not all western cultural products on the TV were adopted. Some were discarded as not suitable to local forms of life and the state's celebration of national ideology. King (1990a) also argues that nation-states constantly aim to construct, define and monitor national cultures within the politically defined boundaries of the state.

Although the decline of the state has been considered as an inevitable result of globalization (Ohmae, 1995), many national and city governments have tried to cushion some of the negative influences of 'the foreign' within their territories. The promotion of ideas like 'the "re-Japanization" or "re-Asianization" of Tokyo' (Machimura, 1998) and 'Seoul Renaissance' (Kim, 1998) by local governments shows a trend of relocalization and renationalization as the oppositional trend to globalization. Re-evaluation of traditional values and heritage has been encouraged by the worldwide spread of postmodern rhetorics in the 1990s. In Seoul, the local government has been extensively involved in (re)inventing tradition, building new monuments and renovating old ones (Kim, 1998).

There has been a growing body of studies trying to go beyond the dichotomy of homogenization and heterogenization of global culture (Featherstone, 1995; Hall, 1997; Hannerz, 1997; Wallerstein, 1997). The writers in this tradition have tried to develop new concepts to theorize increasing cultural flows on a transnational scale and its consequences. Hall (1997) proposes the notion of 'the global post-modern'. He draws attention to the complicated relationship between the US and Latin American countries and asks 'How those forms which are different, which have their own specificity, can nevertheless be repenetrated, absorbed, reshaped, negotiated, without absolutely destroying what is specific and particular to them?' (p. 29). Hall continues that 'The global post-modern is not a unitary regime because it is still in tension within itself with an older, embattled, more corporate, more unitary, more homogeneous conception of its own identity' (p. 32).

Globalization is not a one-way process, but the global is adapted to differentiated local conditions. As the interconnectedness of diverse local cultures grows, differences within societies, rather than between them, have been noted (King, 1997). Accelerating cultural flows between countries, regions and places have created different cultural mixes as more cultures have influenced, and been influenced by, other cultures. In the process of persistent cultural interaction, cultures around the world have more common components, thus reducing extreme differences. At the same time, however, people at the local level are experiencing many different cultures and becoming more familiar with cultural clashes between their own culture and imported ones. Cultural globalization processes are producing the proliferation of differences, the aesthetics of crossover, the aesthetics of the diaspora and the aesthetics of creolization.

The reterritorialization of culture

One important term conceptualizing increasing cultural clashes within a society is the 'reterritorialization of culture'. In order to understand this concept, we need to explain the ideas of territorialization and deterritorialization of culture. In simple terms, these concepts involve a theorization of the connection between place and culture, location and identity.

Some argue that globalization processes have caused a disconnection between original identity and traditional location. Detroit has one of the largest 'Arab' populations outside the Middle East. Melbourne in Australia is the fourth largest 'Greek' city in the world and there are just as many 'Scots' living outside Scotland as inside the national border. The quotation marks indicate the ambiguous nature of these identities. Amin and Thrift (1994) note that the intensification of global communications and international migration has led to the rise of deterritorialized signs, meanings and identities.

The concept of deterritorialization helps us better understand the cultural dimensions of globalization which have been too narrowly defined by the concept of Americanization or homogenization of the world. McDonald's in Moscow can be understood not only as the thorough Americanization of the world, but also simply as a deterritorialized piece of American culture in a post-Socialist city. However, the term is not really suggestive of the presence and, in many cases, the strengthening of heterogeneity around the world. We propose a new concept of reterritorialization to describe the process in which deterritorialized cultures take roots in places away from their traditional locations and origins. The reterritorialization of a culture embraces a series of processes ranging from diffusion from their origin across borders (spatial, temporal and cultural) to establishment in a new place in a new form. Reterritorialized cultures are not simply transposed, they are transformed. McDonald's restaurants in Japan are selling the Teriyaki McBurger that is a sausage patty on a bun with teriyaki sauce (McDonald's Corporation, 1998).

The concept of reterritorialization allows for a better understanding of the strong presence of heterogenizing forces in each society. Reterritorialization is a very powerful concept in the analysis of the process of heterogenization of cultures and the idea of global culture in the plural.

English as a global language

Culture is language as much as language is culture. An important example of cultural de-, re-, territorialization is the spread of English use across the globe. The fast expansion of the English-speaking ecumene has been observed over the past century (Crystal, 1997). The use of English as standard communication code has facilitated globalization processes and has also been strengthened by these processes. In this section we seek to understand the development of English as a global language in the context of cultural reterritorialization, rather than cultural homogenization or heterogenization.

Cultural globalization is intimately connected with the development of English as a global language. English is clearly a hegemonic language in the contemporary world, due to both the extensive impact of the British Empire in the colonial period and the dominance of the American economy, culture, science, technology and politics in the contemporary world.

By 1995, 75 countries officially recognized English as a primary or secondary language. The number of people who speak English as a first or second language has been estimated at approximately 573 million. Contemporary researchers have speculated that a further 670 million may have native-like fluency in the language (Crystal, 1997: 61). It is also true that a large number of non-native speakers as well as native speakers are exposed to the language on a daily basis through advertising, government functions, or interpersonal communications. David Crystal describes them as members of the expanding circle, who have, via the dynamic processes of globalization, been exposed to the usage of English on informal levels. Estimates as to the number vary considerably, but high estimates hover around the one billion mark. Combining this figure with the 573 million first- and second-language speakers, we can estimate that almost 1.6 billion, over a fifth of the world's population, exhibits some reasonable competence in the language.

Language is power and power is language. The dominance of English has accelerated globalization processes, facilitated the dominance of American cultural products over the world, and created increasing needs and desires to learn English. The use of English around the world, however, is not simply a diffusion from a dominant center. There has been a conscious adoption of English around the world. People in non-English speaking countries have been trying to learn English to participate more fully in international activities. Speaking English is required to be competitive in global markets. This explains why so many countries around the world have adopted English as a second language and emphasized it as an important subject in their schools. For example, in the early days of the People's Republic of China (PRC) English was labeled as an 'imperialist language' and its use was suppressed. After the break with the Soviet Union, however, teaching Russian was abandoned and the teaching of English was revived and relabeled as 'the instrument for struggles in the international stage'. In contemporary PRC there has been a great increase in the use of English, fueled by the increasing number of students educated in English-speaking countries, especially the USA, and an increase in the demand for English language skills among professionals and workers in the booming international trade sector. An English-language teaching industry is emerging throughout the country. And even those without spoken ability in English have been subject to the symbols of English-language brand names, such as Coca Cola and McDonald's.

For individuals in many countries around the world, English skills are an invaluable asset in the job market. For example, many multinational Korean companies prefer English-speaking workers because they consider them more adaptable to the global economy. English has become an important form of cultural capital both for national governments seeking to produce a workforce familiar with the communication mode at the global level and for individuals eager to achieve a better position in a globalizing world.

Not only has English expanded to become the dominant worldwide medium of communicative practice, but it also has become a mode of resistance and reinvention by non-English-mother-tongue communities, many of whom are speaking English in their own way, utilizing the plasticity of an accepted norm, speaking the English language while simultaneously bending and purging its hegemonic force to fit their own social identity. To Korean people, for example, someone's 'second wife' means 'mistress', instead of the wife from his second marriage. They use the English term second wife in their everyday life, ignoring its original meaning. English, a global medium for communication, is being reterritorialized within particular communities. There has been a creolization of English in both non-Anglo-American and non-English-speaking countries (Kachru, 1990). We can term it the reterritorialization of English in a culturally globalizing world. The language has not merely been distributed by American economic and cultural forces, but also adopted, promoted and creolized.

6

Cultural Globalization and the City

Global ghettoes, villages, nations, alienations, and, most important, global communalities and communities have become possible as never before.

(Ingram, Bouthillette and Retter, 1997: 3)

In this chapter, we explore the relationship between cultural globalization and the city in three ways. First, we consider shifting identity in a globalizing world. The state has been a strong organizational cultural force, constructing people's identities by citizenship. As nation-states fade in economic significance and more people claim multiple nationalities, globality and locality, as much as nationality, have become important sources of identity. Second, we consider the relationship between the global and the urban in more detail through an examination of ethnic identities. Third, we look at the rise of the cultural economy. Culture has come to be a commodity. Cities around the world are fully aware that they can boost their economies through entertainment industries and urban tourism. The commodification of culture is an important element in urban economic health. We consider the role of spectacle in the symbolic economy of the city.

Identity, culture and the city

Globalization involves the creation of global identities. Shared consumption patterns, similar work experiences and a worldwide circulating repertoire of images and icons are important elements in the shaping of these identities among people. The formulating process of global identities, cul-

tural globalization in a broader term, is dialectic: as people become more global they also become more national, more ethnic, at least in certain parts and times of their lives. Young Japanese couples often have two wedding ceremonies: a traditional one and a more Western-style wedding often in a Christian church. One Episcopal diocese in Australia, popular with Japanese wedding groups, struggles with the problem of wanting revenue from conducting such ceremonies but wary of partaking in an obvious commodification of a sanctified ceremony.

It is not a case that, as global connections increase, people lose their sense of difference; if anything globalization can make groups feel even more distinctive as they now have a wide range of samplings of other cultures. This difference is often promoted by the tourist gaze which seeks national difference, cultural sampling and 'authenticity'. As the world becomes more the same, difference is prized, visited, created, commodified. Cities become settings for national and group displays of culture in museums, buildings, festivals and spectacles.

Two of the most defining elements of identity in the nineteenth and early twentieth centuries were nationality and class. Both of these have become more problematic sources in a globalizing world.

NATIONAL IDENTITY IN CRISIS

The nation-state, one of the most important determinants of group identity, has been weakened. National borders are becoming so porous, at least for some groups, that they no longer fulfil their traditional role as barriers to the movement of goods, finances, ideas and people, and as a marker of the extent and power of the state (Ohmae, 1995). As a result, the role of certain individuals in these structures is called into question, especially in terms of their loyalties and identities. The fall-off in the determinative power of states has given rise to the new politics of identity, in which the definitions of citizenship, nation and state vie with identities which have acquired a new political significance, such as gender, sexuality, ethnicity and race, among others, for control of the popular and scholarly political imaginations of the contemporary world (Wilson and Donnan, 1998). According to Mercer (1991, 43), identity becomes an issue only when it is in crisis, when something assumed to be fixed, coherent and stable is displaced by the experience of doubt and uncertainty. Identities are multiple, frequently shifting and often contradictory. There is a plurality of identities and identity construction is an ongoing, ever changing part of one's life in the global world. The young golf champion Tiger Woods embodies the multiplicities of identity; he is black with Thai heritage from his mother's side and Native American from his father's side. There is no single appropriate category on the US Census form for his ethnicity.

Multiplicity, discrepancy and liminality of identities is growing more obvious in a globalizing world. The demise of a meta-, nation-based, all-embracing identity of citizenship has given rise to other identities which are more flexible and ambiguous. Garcia (1996) points out two main components shaping the vulnerability of the citizenship principle in the western world: the retrenchment of the welfare state and the impact of immigration. Citizenship has been associated with membership of a state, nationality. This membership has been defined as the combination of certain rights and obligations in political, economic and social terms. While political rights of citizenship have consistently been exercised in most economically developed, politically democratic countries, rights to economic welfare and security for the poor are currently at stake. The growing number of Turkish 'guest workers' in Germany, of Mexican illegal immigrants in California and of Eastern Europeans in Western

European cities is creating changes in the concept of citizenship. Most of these migrated people are effectively second-class citizens who lack access to social services, and experience political exclusion and economic disadvantages.

Many nation-states, however, are resisting the gradual denationalization of citizenship. Along with sovereignty and exclusive territoriality, citizenship is still a critical institution of the modern nation-state. Many states still maintain a policy that forbids dual citizenship. While numbers of (il)legal immigrants are growing fast, many countries, particularly developed countries, have been trying to keep the door closed to illegals. Mitchell and Russell (1996) note that Western European governments have made concerted efforts in recent years to develop new and tougher forms of immigration control. The political and social status of illegal immigrants, mainly from Mexico and the Caribbean, has been jeopardized by the Republican-leading US Congress over the past few years. Sassen (1996: 159) notes that economic globalization denationalizes national economies, while in contrast immigration is renationalizing politics. There is a growing consensus among states to lift border controls for the flow of capital, information and services and, more broadly, to further globalization. But when it comes to immigration and refugees, whether in North America, Western Europe or Japan, states claim their sovereign rights to control borders. Following the creation of the Single Market in Europe, labor migration has been spurred across the European Union (Jenkins and Sofos, 1996). Citizenship of the EU, however, can only be acquired by individuals holding citizenship in one of the member states, even though millions of other residents have been living in Europe for more than a generation (Garcia, 1996; Kofman, 1995; Mitchell and Russell, 1996). EU citizenship is entirely dependent on a traditional conception of national citizenship. Nationality and citizenship are still stubborn facts of political life that resist the easy adoption of global identity.

CONSUMPTION AND IDENTITY

For Marx, identity and political consciousness was primarily a function of production. The central dichotomy was whether you owned or did not own the means of production. Today, however, the consumption of goods is at least as important as the production of goods as a central defining element in individual and group identity. Fashion consciousness has replaced class consciousness. Although Marx proposed the coalition of the working class around the world through the famous claim 'Workers of all countries, unite', class as an agent of history, as E. P. Thompson would use it, has been defined mainly within a national society. His classic work, first published in 1965, is entitled *The Making of the English Working Class*. The national qualifier is a major component of class. The making of a working class with global consciousness has yet to be achieved.

Indeed, it has been argued that the traditional working class within a society has been socially fragmented and damaged by ethnicity, gender and location in the labor and consumer market (Fainstein *et al.*, 1992). The dominance of class as the master, unitary identity has been undermined by the growth of various new social movements, such as feminism, human rights and environmental movements, among others (du Gay, 1996). These movements are often operating on a transnational scale and in the case of human rights they have an international surveillance.

The rise of service industries in the developed economy has also shaken the identity of a modern economy structured in the image of manufacturing. Since the overwhelming bulk of employment in most developed countries is currently generated by service sectors, class consciousness strictly associated with manufacturing employment is no longer such a defining element in identity creation. Rose (1990: 102) notes that the primary economic image offered to the modern citizen is not that of the producer but that of the consumer. He continues, 'People, as consumers, are encouraged to shape their lives by the use of their purchasing power and to make sense of their existence by exercising their freedom to choose in a market. Thus in a consumer, enterprise culture, freedom and independence emanate not from civil rights but from individual choices exercised in the market. Students, patients and parents have all been re-imaged as customers.' When a private col-

lege promotes credit cards to its students, the distinction between students and consumers becomes blurred.

There is an increasing connection between consumption and identity, most clearly evident in the growing focus on certain lifestyles in advertisements. Adverts reflect and create group identities. In *Distinction*, Bourdieu examines how consumption practices in the fields of the arts, food, interior decor, clothes, popular culture, hobbies and sports play a role in the making of people's status, particularly the cultural elite's class position (Holt, 1998). The consumption of certain products, such as luxury cars, has been aesthetized as the expression of distinct tastes as well as wealth. In a market-dominating consumer society, notes du Gay (1996: 100), those whose consumption does not matter much for the successful reproduction of capital are not *represented* in the many different meanings of the word.

The progressive penetration of the market into our everyday life has led to the appearance and dominance of consumption-based identities. In the literature there have been continual calls for a need for a more grounded understanding of the links between consumption and identity (Crewe and Lowe, 1995; Jackson and Holbrook, 1995; Jackson and Thrift, 1995). Theorists have argued for a rapprochement between economic and cultural approaches, a tighter meshing of production-oriented and culturally derived understandings of consumption and identity, and a more empirically grounded, historically sensitive and geographically contextualized style of research to understand the cultural politics of consumption (Jackson, 1995).

GLOBAL IDENTITY AND THE CITY

O'Connor (1998) identifies the emergence of new forms of cultural consumption and the construction of lifestyle in the contemporary city: a dramatic increase in the production and consumption of symbolic goods; the shift of consumption from use value to sign value; the destabilization of established symbolic hierarchies through the articulation of alternative tastes and styles; the rise of popular and commercial cultures as alternative forms challenging established high culture; and

Table 6.1 Where the transnational business class prefer to stay

(a) Top 10 US city hotels

Rank	Hotels	City
1	Four Seasons	Chicago
2	The St. Regis	New York
3	Hotel Bel-Air	Los Angeles
4	Four Seasons	New York
5	Ritz-Carlton	San Francisco
6	The Peninsula	Beverly Hills
7	Mansion on Turtle Creek	Dallas
8	Mandarin Oriental	San Francisco
9	Ritz-Carlton	Chicago
10	The Lowell	New York

(b) Top 10 international city hotels

Rank	Hotels	City
1	The Peninsula	Hong Kong
2	The Regent	Hong Kong
3	The Oriental	Bangkok
4	Hotel Ritz	Paris
5	The Stafford	London
6	The Lanesborough	London
7	Le Bristol	Paris
8	The Connaught	London
9	Hotel Cipriani	Venice
10	Four Seasons	Milan

Source: Forbes (http://www.forbes.com/forbes/97/0922/6006278a.htm)

the emergence of new urban spaces creating 'play spaces' for new forms of sociability.

Global identities are reinforced by the shared consumption of similar goods and images. We can consider different types of global identities. The most obvious is a transnational business class. This group operates in a global context, traveling easily between countries; they stay in the same hotels and draw upon the icons of global *savoir-faire* drawn from the around the world – Italian clothes, French champagne, US popular culture, ethnic food. There is a repertoire of consumption that can be found in business elites around the world.

Table 6.1(a) and (b) lists the world's and the US's best hotels as cited in the in-house magazine of the business class, *Forbes* magazine. Top hotels are located in New York, Chicago, Hong Kong, London and Paris (*Forbes*, 1997). This group can be termed the transnational business class to refer to their shared sense of identity and, using a popular category from airline travel, their mode of moving around the world. This class sits atop the global economy and, while there are national differences, what is apparent in the last two decades is how these differences are lessening. A global business class is not defined simply by a shared set of consumption goods or mode of travel; it is a self-conscious group connected by constant inter-action and shared interests. A global economy is tied together not just by flows of goods and ser-vices but by social interaction between an increasingly tight-knit class. This class invariably live or work in, or just visit, world cities. Their shared identity is reflected in the luxury shopping areas, the business class hotels, the VIP lounges in airports.

Sklair (1996) examines the formation of the transnational capitalist (business) class in Australia. He identifies four distinguished groups comprising the class: transnational corporation executives and their local affiliates; globalizing bureaucrats; capitalist-inspired politicians and professionals; and consumerist elite (merchants and media). Sklair argues that these Australian-based upper-class people are transnational in that they tend to have internationally oriented per-spectives on a variety of issues; they are from many countries around the world; they increas-ingly consider themselves global citizens; and they tend to share similar lifestyles, particularly patterns of luxury consumption of goods and ser-vices. Clearly, the transnational business class facilitates globalization processes through their global knowledge, global sourcing and global movement.

In deep contrast is the experience of a global working class. If we take 1970 as a base point then the developed world still undertook the bulk of global manufacturing. A long-established working class had developed in North America and Western Europe. Skilled, predominantly male workers formed the basis of the class, who had achieved tremendous strides in improving living and working conditions. A Keynesian compromise had been reached. The global shift of manufactur-ing has created more manufacturing workers in the world, but there has been a regendering and reterritorialization of the working class. More manufacturing jobs are being undertaken in devel-oping countries by young women. The traditional working class has been restructured. Workers in the developed world have seen mounting pressures

on wages and working conditions while many workers in the Third World have seen an improve-ment in their living conditions. Unlike the business class the working class is more fractured, less globalized in connections and mutual understand-ing. Marx had two definitions of class – class in itself and class for itself. A global working class has been created in itself, but a class for itself, self-consciously aware of its mutual interdependence, has yet to develop at a global level. The business class is thoroughly internationalized, the working class is still more national.

The terms business and working class are broad-based definitions. A more detailed examin-ation of similar groups across the world or of different groups within one city would reveal a more complex pattern. At a broad level Reich (1991) identifies those working as symbolic ana-lysts and the winners in the global economy. Connected to booming global economies this group, while living in the same city and nation as say manufacturing workers in declining industries, are experiencing very different life chances. At a more detailed level, Mead (1998) identifies differ-ences among symbolic analysts in US cities: winners include software programmers and enter-tainers; those whose outlook may be bleaker include professors, accountants, doctors and lawyers.

Between the business class and the working class there has emerged a large middle group whose identity is shaped more by consumption than by production. This amorphous mass con-nects with culture more than the economy for sources of individual and group identity. The global reach of specific goods and services such as Nike has led to a market homogenization. There has been a counter-trend: market segmentation that reflects differences in income, gender and especially age rather than nationality. The world is not becoming the same, but certain parts of it are becoming more alike (Figure 6.1). Youth culture in its various forms, for example, has become more internationalized. The global popularity of Leonardo di Caprio and the Spice Girls, for example, is just one element in a teen culture that transcends national boundaries. Young middle-class people in rich countries around the world have a similar repertoire of clothes, idols, songs

Figure 6.1 The coffee-shop culture of world cities: a Starbucks in Chicago. (*Photograph by John Rennie Short*)

and computer games to draw upon for their commodified tastes and symbolic appeal.

GLOBAL IDENTITIES AND GLOBAL CITIES

I noted the list of stores along a hundred-yard section of Parnell Road in Auckland, New Zealand, in 1998 (Figure 6.2). They included an Italian bakery, a fine art gallery, a wine merchant, a French antique furniture store next to a shop selling a range of goods from Indonesia, a German pastry shop, DNKY and Timberland clothing stores, a furniture store where Le Corbusier couches sit beside copies of the chairs of Charles Rennie Macintosh and Mies van der Rohe; and a wide selection of cafés and restaurants, with names such as *Metropole* and *Milano Ristorante*, where people, dressed predominantly in black and invariably slim, could sample cuisine from Italy, Japan and the southwest US.

In world cities there is a cosmopolitan feel. The term cosmopolitan comes from the Greek *kosmos*, meaning world. Cosmopolitan means pertaining to the world or free from national limitations. World cities tend to be cosmopolitan. The freedom from national limitations can take a number of forms. First there is the cos-

mopolitanism associated with a wide range of consumption goods and associated identities. The Parnell Road area is a good example. Second, there is the cosmopolitanism that comes from a wide range of racial and ethnic identities within a city. Global cities have peoples and communities from all over the world. In an influential book the geographer Ed Soja (1989) asserted that 'It all comes together in Los Angeles.' The remark is typical Californian boosterism with tinges of a hidden anxiety that it may not be true. Los Angeles is a world city, if not quite *the* world city its proponents fantasize.

Not all world cities are equally cosmopolitan. Tokyo (Figure 3.1(b)) is still more Japanese city than world city. Montreal may not be called an important globalizer in terms of the world economy, but the city's restaurants, ranging from Lebanese to Ethiopian, are fairly cosmopolitan. There are in Montreal eight bookstores, entitled *Maison de la Presse Internationale*, for foreign books, magazines and newspapers serving the city's cosmopolitan readers.

Nijman (1996) makes a distinction between world city and cosmopolitan city. He argues that while Miami holds a minor position in the international urban hierarchy as measured by conventional measures of command functions, it

Figure 6.2 Billboard on Parnell Road, Auckland, New Zealand. (*Photograph by John Rennie Short*)

is more of a world city in terms of cosmopolitan culture with both a Caribbean and a Latin American flavor to the South Florida metropolis. Boniche (1998) examines the construction of a global culture in the city. He shows how going topless on public beaches was suggested in 1985 by a former mayor, to solidify Miami Beach's appeal as an international tourist destination. Going topless was a cyclical phenomenon restricted to the tourist season but each year local beachgoers adopted the customs of the international tourists. A global culture was adopted, making the city less a provincial North American and more a global place.

Cosmopolitanism, at least in terms of consumption, has become such a characteristic of world cities that cities seeking to proclaim their city status emphasize their range of urban experiences. Cities now boast of their range of cuisines, ethnic festivals, and mix of races and ethnicities. Cities that lack the cosmopolitan feel are considered provincial, narrowly national in an increasingly global world. An important function of city reimaging, by both public and private agencies, is to foster the sense of a cosmopolitan city rather than a regional or national city.

Ethnic identities and the global city

The distinction between global and local is no longer an easy one to make. Beauregard (1995) describes the relationship between the two scales as dialectical. This is most apparent when we consider the seeming paradox of ethnic globalization. Ethnicity and globalization are often seen as different, reflecting structural incompatibilities such as local–world, place–bound/space-free. We argue, in contrast, that a global world has often strengthened ethnic ties and identities across the globe rather than weakened them. Ethnic globalization has created dual and multiple identities as diasporas are more easily connected through time and across space. Ethnicity is easier to maintain because of the commodification of culture, the politics of identity, and the technological ease of maintaining diasporas.

In 1997, more than a third of New York City's population (36.1%) were foreign-born (Figure 6.3). According to a series of articles in *The New York Times* (Dugger, 1998; Sontag and Dugger, 1998; Sontag, 1998), these modern immigrants are different from their predecessors early in this century who abandoned their motherlands forever, shutting one door, opening another and

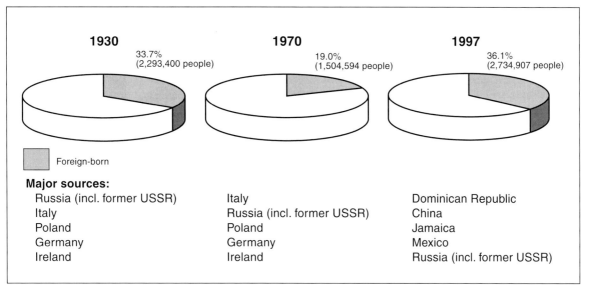

1930	1970	1997
33.7% (2,293,400 people)	19.0% (1,504,594 people)	36.1% (2,734,907 people)

Foreign-born

Major sources:

Russia (incl. former USSR)	Italy	Dominican Republic
Italy	Russia (incl. former USSR)	China
Poland	Poland	Jamaica
Germany	Germany	Mexico
Ireland	Ireland	Russia (incl. former USSR)

Figure 6.3 Immigrant city: foreign-born in New York City. (*Source:* Sontag and Dugger, 1998: 28)

never looking back. The new immigrants are more tightly bound to their homelands than ever before. Their wealth that allows them to visit relatives in homelands is one factor explaining this new immigrant experience. Of course, advances in technology have reduced the cost of connecting the two worlds and created new easy ways to telecommunicate, such as e-mail and videophone. The growing intensity of their business ties to their old countries is another important factor, reflecting the blossoming of international economic flows: the money-wiring business between Santo Domingo and New York City carries $500 million a year and the gem trade between Jaipur in India and New York City involves 230 trading companies owned by Indian immigrants.

Regular visits and strong economic ties to their homelands have led these immigrants to have dual identities and dual loyalties. Although many migrant workers have suffered, and some continue to suffer, economic difficulties, cultural clashes and political discrimination, it is not necessary to picture their identity and existence in terms of crisis, displacement and disjuncture. Their ambiguous belonging may rather allow them to make the best of their multiple memberships: one, New York City where they are currently living; the other, their hometown in India, the Dominican

Republic, Mexico and other points around the world.

Ethnic areas in world cities can also function as transnational business centers, anchor points firmly embedded in international flows of people, ideas and capital, and attractive tourist sites. These immigrant economies are an active part in economic and cultural globalization (Figure 6.4). The ethnic network has played a crucial role in the Chinese ethnic economy in Los Angeles, which facilitates US–Chinese trading patterns. The network operates in a unique way reflecting not just Chinese inherited business practices but also learned practices from American capitalists. According to Razin and Light (1998), ethnic entrepreneurs in large metropolises in the US have formed their own ethnic economies regardless of metropolitan opportunity structure. Entrepreneurs having small businesses in their ethnic communities are not heavily affected by local downturn. Local ethnic economies may be more tuned to the global economy than to the national one.

Ethnic identity is not a fixed, stable category. It has always been partly adopted, partly imposed, always in the process of re-creation. According to Foner (1997), most immigrants in the US create new kinds of family patterns which are different from those in their home countries and from

Figure 6.4 A 'Chinese' economic district in Auckland, New Zealand. (*Photograph by John Rennie Short*)

others in the US. New immigrant family patterns are shaped by cultural meanings and social practices that the immigrants brought with them from their old home as well as by social, economic and cultural forces in their new home. In the global world it has become possible to maintain multiple identities. When a Scotsman and a Korean live in America, are they Scots-American, Korean-American, or do they become expatriate Scots and

Korean and some blend of all of them? Ethnic globalization is a process of flux and change with both origin and destination, home and abroad losing their firm categorization.

Ethnicity is both imposed and created; it is a social construction rather than a biological fact. Ethnic identities are being reconstructed in many cities around the world. There is the construction of Chinatowns in many cities, both reflecting a

Chinese presence, but also part of the drive to give the feel of the cosmopolitan city so vital to achieve and maintain world city status. The presence of different ethnic areas, especially for eating and shopping experiences, has become an important indicator of world city status.

Urban cultural economics

There has been a tendency to focus on economic readings of global spaces, most notably the processes of investment, speculation and capital switching underpinning the rise of global cities, and to neglect the cultural determinants of uneven development. Many argue against research which takes its cue only from political economy and treats consumption only as an aspect of the circulation of capital (Crewe and Lowe, 1995; Lash and Urry, 1994). Such analyses must move on from being one-sidedly productionist, and must grasp the extent to which culture, aesthetics and symbolic processes have penetrated the economy.

Crewe and Lowe (1995) advocate a combined analysis of material process and symbolic representations focusing on the spatial outcomes of the complex mediation between retailers, advertisers and consumers of the fashion industry. Through a case study of Nottingham's lace market, they examine the ways in which retailing and consumption create and recreate place-specific identities, and how cultural attributes and capital gain are intrinsically bound together. They conclude that increasingly, what distinguishes one place from another is the strength of their consumption identities, also indicating the emergence and significance of distinctive new urban landscapes.

The blurred distinctions between the economic and the cultural, production and consumption, have been noted by urban policymakers and academics (Kearns and Philo, 1993; King, 1996; O'Connor and Wynne, 1996; Wynne and O'Connor, 1998; Scott, 1997; Zukin, 1991, 1995). Some characteristic cultural traits, including historic sites, exotic landscapes and art streets, have been major destinations of tourists, functioning as important economic generators in many cities. There is, however, a new trend that most cities are now beginning to utilize culture for economic gains, bringing together cultural and economic policy at an ever more strategic level (Kearns and Philo, 1993). Du Gay (1996) considers the rise of the cultural industries as part of the gradual de-differentiation between economy and culture in late capitalist society.

The increasing importance of culture in the city can be identified in three ways: the rapid rise of aesthetic, cultural and symbolic landscapes in the city, the growing contribution of cultural industries to the urban economy, and the importance of spectacle.

THE SYMBOLIC ECONOMY OF THE CITY

The utilization of the cultural as a significant redevelopment strategy has repeatedly been argued by Zukin (1982, 1991, 1992, 1995). She has developed the notion of the symbolic economy of the city consisting of the symbolic languages of exclusion and entitlement involved in the use of aesthetic power by place entrepreneurs. 'City advocates and business elites', she argues, 'through a combination of philanthropy, civic pride and desire to establish their identity as a patrician class, build the majestic art museums, parks, and architectural complexes that represent a world-class city' (Zukin, 1995: 7–8).

Facing various forms of urban distress in the 1970s, many cities have undergone urban redevelopment. Many derelict areas in the central city have been cleaned up under the name of revitalization, regeneration and gentrification. Old buildings have been torn down. Industrial complexes have been pushed out. Office buildings, museums, galleries, upper-division retail shops and high-class restaurants have emerged in downtown New York City, London, Boston, Philadelphia and many other cities. Zukin (1991) attributes the remarkable aesthetization of city centers, accompanied by postmodern-looking office buildings, the rise of up-market consumption and inputs from high cultures, to a form of culturally based urban renewal.

The aesthetization and culturalization of the city can be connected to the manipulation of city images and city marketing. Cities are placing great emphasis on their looks. Numerous central city revitalization programs have significantly transformed many large metropolises into such glitzy places. Cities are striving for jumbo-sized sports stadiums which dominate urban skylines. Postmodern buildings, ethnic streets, theme parks, museums and cultural centers have become the most important visual representation of many cities.

The look and feel of cities reflect decisions about what – and who – should be visible and what should not. Through an analysis of cultural texts and visual images on the growth of Las Vegas and Coney Island's decline, Zukin *et al.* (1998) argue for the clear resonance between a city's ability, or inability, to attract residents, tourists and investment capital, and its representation of significant social themes.

According to Zukin (1995), the enormous efforts of many cities to enhance the visual appeal of urban spaces has led to a significant number of new public spaces owing their particular shape and form to the intertwining of cultural symbols and entrepreneurial capital. The display of art, for public improvement or private gain, represents an abstraction of economic and social power. Among the business elite, those from finance, insurance and real estate are generally great patrons of both art museums and public art, as if to emphasize their prominence in the city's symbolic economy.

Creating a public culture involves both shaping public space for social interaction and constructing a visual representation of the city. The proliferation of public–private partnerships in urban (re)development processes has accompanied the strong involvement of private-sector elites, both individual entrepreneurs and big corporations, in the remaking of urban space and public culture. The participation of private interests in the building of public space often changes the nature of public space. The turbulent building history of Monona Terrace in Madison, Wisconsin, for example, shows how art, politics and the market are entangled with one another (Puhl, 1998). The convention center, originally designed by the architect Frank Lloyd Wright as a civic center in 1938, was built and opened in July 1997 in pursuit of boosting Madison's sophisticated, high-culture image. Unlike Wright's first draft of the building, Monona Terrace does not function as a 'truly' public space, but as a convention center for urban tourists and professional conventioneers. The business and government coalition of Madison has exploited Wright's name, design and dream of a true civic center to produce a convention center. Sophisticated image making is considered to be critical to urban competitiveness and growth. The public–private coalition of Madison uses the Wright mystique to sell the city.

THE CITY OF CULTURE

Cultural policy within cities had traditionally largely centered on the maintenance of the established institutions of civic culture – the public library, the civic center for art and performance, the civic square. Contemporary cities, instead, are beginning to embrace leisure, entertainment and atypical lifestyles into their political discourse. There is a magazine advertisement, entitled *California: Culture's Edge*, which Los Angeles, San Diego and San Francisco have inscribed jointly. The ad shows that these cities consider hippie culture part of their lifestyle. Furthermore, they are trying to attract tourists through their 'characteristic' culture. On the following page of the same magazine, San Diego pictures itself as a city with 'class' including historic architecture, theater, museums and opera. Today's large metropolises advertise culture in all its forms, from high culture through pop culture to gay culture. Sydney now actively markets its Gay Mardi Gras (see below).

Cultural promotion has become closely tied to city promotion. Culture is playing a significant part in the refashioning of collective emotion and consciousness within cities. New forms of civic identity are being shaped through investment in cultural facilities (Boyle, 1997). The growth of disposable income – especially among the youth – has created huge growth in the scale and scope of the cultural industries and their associated products.

There is a connection between cultural politics, cultural economics and the selling of the city in the contemporary world. One example is provided by the emergence of the gay community in many world cities in recent years. A number of studies have outlined the emergence of a gay culture in cities. George Chauncy (1994), in his careful examination of the gay male world in New York City from 1890 to 1940, undermines the myths of isolation and invisibility. He presents a picture of a community with relative cultural autonomy and a high degree of visibility both to the outside world and to each other. This world has been so long forgotten, it is almost believed that it never existed. In more provincial cities and even in New York City after the Second World War, gay communities were more marginalized. John Grube (1997), for example, shows how the male gay community in Toronto was activated by police raids on bathhouses in 1981. Prior to this, many gays hid their identity and public space was used surreptitiously rather than openly. In the aftermath of the general politicization after the Stonewall resistance in New York City in 1969 and the Toronto police raids of 1981, a more open community developed in which male gayness became a source of identity rather than a source of shame.

The development of a more open gay identity and the emergence of 'gay' areas and cities has become tied to city image making. In the 1980s Manchester in England emerged as a major center for gay culture. The city had a self-proclaimed 'gay village'. In an interesting study Stephen Quilley (1997: 275) shows how 'the gay presence in the city has, since 1987, been increasingly incorporated into a postmodern narrative of cosmopolitan diversity ... the aesthetic of the gay scene has become articulated into a wider reimaging of the city around the familiar theme of European style cafes, pedestrian streets, and arcades, as well as around a central role for leisure and cultural activities.'

Even gay politics as resistance can end up as commodified images of a promoted city. The Gay Mardi Gras in Sydney began as a celebration of gay pride. On 24 June 1978 almost 1,500 people took part in a protest march which ended with a confrontation with the police. The next year saw another Mardi Gras, almost 5,000 taking part. It became an annual event where outrageous cos-

tumes were proudly worn, less of a protest and more of a celebration. By the early 1980s the march had become a parade. In 1983 there were 44 floats watched by 20,000 spectators. By 1994 there were 137 floats watched by 600,000 spectators and a large TV audience who were able to watch edited highlights on network television. The parade became one of the official festivals of the city. What had begun as an act of resistance had become part of the official image of the city, used to reflect its cosmopolitan, multicultural character. The global city needs this character. The Gay Mardi Gras became an important source of revenue and civic pride.

THE CITY AND SPECTACLE

Spectacles have often been used as urban promotional events. Cities can increasingly be seen as commodities whose marketing is aided by the hosting of spectacles, public events and major athletic competitions.

The relationship between cities and spectacles is not a recent event. The most significant in the modern era – indeed it helped to *define* the modern era – was The Great Exhibition held in London in 1851. Its full title was 'The Great Exhibition of the Industry of All Nations'. It involved the construction of the Crystal Palace, designed by Joseph Paxton, whose arches of glass and iron were to reappear in the postmodern sensibilities of the 1970s and 1980s. The Palace covered 17 acres. At the heart of the exhibition was the display of commodities. There were over 8 miles of display tables. Manufactured objects were carefully arranged and displayed. Charlotte Brontë was just one of many visitors:

> Yesterday I went for the second time to the Crystal Palace. We remained in it about three hours, and I must say I was more struck with it on this occasion than at my first visit. It is a wonderful place – vast, strange, new, and impossible to describe. Its grandeur does not consist of *one* thing, but in the unique assemblage of *all* things. Whatever human industry has created you find there, from the great compartments filled with railway engines and boilers, with mill machinery in full work, with splendid carriages of all kinds, with harness of every description, to the glass-

covered and velvet-spread stands loaded with the most gorgeous work of the goldsmith and silversmith ...
(Quoted in Carey, 1987: 324)

The Great Exhibition displayed the artefacts of the capitalist bourgeoisie order in the capital city of the leading industrial country in the world. It placed the commodity at the center of cultural life, introduced new forms of display and selling, and laid the foundation for subsequent exhibitions and trade fairs.

The 1893 World's Columbian Exhibition was held in the US to celebrate the 400th anniversary of Columbus's discovery of America. Four cities competed to be the venue – Chicago, New York, St Louis and Washington DC. Chicago business interests, eager to attract the event to generate publicity and revenue for the city, put up $5 million and offered to double it if their city was chosen. Congress voted to accept the Chicago offer. The Exhibition was meant to celebrate American business with an emphasis on technology. The local elites were also eager to promote the idea that Chicago was 'back in business' after the disastrous fire of 1871. The Exhibition attracted 28 million visitors and brought in $14 million. The landscape of the city was transformed by beautification schemes and the site itself which created two parks from swamp land. The Exhibition also saw the early formulation and practice of what came to be known as the City Beautiful Movement.

These two examples highlight the importance of exhibitions to both cultural changes and urban transformation. In the twentieth century, while fairs and exhibitions were still important events, the Olympic Games became one of the most important global spectacles. The modern Olympic Games began in Athens in 1896. They were promoted by Baron Pierre de Coubertin as a way to foster better international understanding at a time of deepening national rivalry. One plan was to host them permanently in Greece, the site of the original Olympic Games. Instead they became a moving spectacle competed for by major cities and wannabe cities. In 1924 the Winter Games became a separate entity. The Summer Games have become one of the biggest global spectacles where national rivalries bring together a global television

Table 6.2 Host cities of the Olympic Games, 1896–1960

Year	Host city	Year	Host city
1896	Athens	1928	Amsterdam
1900	Paris	1932	Los Angeles
1904	St Louis	1936	Berlin
1906	Athens	1948	London
1908	London	1952	Helsinki
1912	Stockholm	1956	Melbourne
1920	Antwerp	1960	Rome
1924	Paris		

Source: Monaco, 1998, Olympic Almanac (http://www.andrew.cmu.edu/~mmdg/Almanac/)

audience. The Summer Games are a major opportunity to reimagine and transform a city. Table 6.2 lists the host cities. The list is revealing. There is a strong European and North American bias. London and Los Angeles have held the Games twice. Tokyo was the first Asian city to host the Games, in 1964.

The Games are one of the single greatest marketing opportunities for a city and its host country. Hitler used the 1936 Games to display Nazi ideology. Munich used the 1972 Games to show the 'new' Germany as a legitimate member of the world community.

The sheer size of the Games has become enormous. There were only 13 participating nations in the 1896 Games in Athens. A hundred years later, at the 1996 Games in Atlanta, there were 197 participating nations, 2 million visitors and an estimated television audience of 3.5 billion in a 16-day period between 19 July and 4 August. The Atlanta Committee for the Olympic Games (ACOG) spent $500 million on infrastructure improvement and the construction of new facilities. The Games transformed the city's landscape. The Centennial Olympic Park was constructed on a 21-acre downtown site, a $209 million Olympic stadium was built, now the home of the Atlanta Braves, and major improvements were made to the airport. The city claimed to have made $3.4 billion from hosting the Games. The Games also helped to raise the profile of the city and to con-

Table 6.3 Host and candidate cities for the Summer Olympic Games, 1960–2012

Year	Host city	Bid cities[a]
1960	Rome	Brussels, Budapest, Detroit, Lausanne, Mexico City, Tokyo
1964	Tokyo	Brussels, Detroit, Vienna
1968	Mexico City	Buenos Aires, Detroit, Lyon
1972	Munich	Detroit, Madrid, Montreal
1976	Montreal	Los Angeles, Moscow
1980	Moscow	Los Angeles
1984	Los Angeles	–
1988	Seoul	Nagoya
1992	Barcelona	Amsterdam, Belgrade, Birmingham, Brisbane, Paris
1996	Atlanta	Athens, Belgrade, Manchester, Melbourne, Toronto
2000	Sydney	Beijing, Berlin, Istanbul, Manchester
2004	Athens	Buenos Aires, Cape Town, Rome, Stockholm
2008	?	Boston, Chicago, Osaka, Rio de Janeiro, Toronto
2012	?	Pittsburgh, Seattle, Warsaw

[a] Bid cities, 1960–1972; finalists, 1976–2004; and potential candidates, among many others, which already began various Olympic campaigns, 2008–2012

Source: Monaco, 1998, Olympic Almanac (http://www.andrew.cmu.edu/~mmdg/Almanac/)

firm the official image of a world-class city (Rutheiser, 1996).

Despite the official hype, a question arises of what are the long-term effects on a city. Cameron Stewart (1998) argues that the Games did not become the much hoped-for seminal moment in Atlanta's history. He quotes one resident as saying 'They were great but they are over now and things are pretty much the same around here'. While there is the legacy of the stadium and Centennial Park, and no debt, inward investment did not increase substantially and the downtown, while receiving windfall gains, is still a place not frequented by ordinary citizens. The social fabric has remained untouched by the Games. Ean Higgins (1998) paints a more positive picture of a Barcelona substantially improved by both the infrastructure improvement and the opening up of the waterfront to pedestrian access. All the Olympic facilities are put to very good use; the Olympic village is now an inner-city seaside suburb, and tourism and investment have increased. There is a price, however: taxes in the

city are high, in part to pay off the debt incurred in hosting the Games.

Table 6.2 shows only the cities hosting the Games. Just as revealing is a list of cities competing to host the Games. There is a place war to host the Games to increase a city's visibility, enhance its image and increase its competitive advantage in the global economy (Roulac, 1993). Wannabe world cities compete for command functions and world spectacles. A good, though not infallible, guide to wannabes is to note those cities that have either hosted or applied to host the Summer Olympic Games. Table 6.3 lists such cities since 1960.

The South Korean Sports Minister Lee Yeung Ho noted in 1988, 'Hosting the Olympics gives us international recognition and a psychological boost for our next step to join the advanced countries within the next decade. Look what happened to Japan after the 1964 Olympics' (Maltby, 1989: 206). What happened to Tokyo was not simply the result of the Olympic Games, but the Games aided the internationalization of the city both internally and externally. The Olympic Games are not just an opportunity to be the site of a global spectacle, and hence confer international name recognition; they also provide enormous opportunities for business and real estate deals. Hosting the Summer Olympics is a giant urban redevelopment opportunity that boosters hope can propel a city up world city rankings.

7

Representing Cities in a Global World

Culture has become as necessary an adornment and advertisement for a city today as pavements and bank-clearances. It's Culture, in theaters and art galleries and so on, that brings thousands of visitors to New York every year and, to be frank, for all our splendid attainments we haven't yet got the Culture of a New York or Chicago or Boston – or at least we don't get the credit for it. The thing to do then, as a live bunch of go-getters, is to capitalize Culture; to go right out and grab it.

(Lewis, 1922: 261)

This quote is taken from the 1922 novel *Babbitt* by Sinclair Lewis in which he describes a meeting of Zenith Boosters Club. The city of Zenith, set in the American Midwest with a population of just over a quarter of a million, is the backdrop of the novel. The main characters are local business folk. They spend much of their time boosting Zenith. The novel draws upon a significant feature of civic culture; urban boosterism has a long tradition in the USA and increasingly has become part of the civic culture of a globalizing world. In this chapter we want to place this boosterism in a broader theoretical setting by connecting the issues of representation, boosterism, and the notion of capitalizing culture to the broader processes of economic and cultural globalization.

Representing the city

Space is turned into place through acts of discursive representation. The generality of space is turned into the particularities of place through acts of description and evaluation. We will use the term *urban representation* to describe these acts.

Cities can be usefully seen as acts and embodiments of representation. Cities are represented in a variety of ways: street maps, coats of arms, sports teams, slogans, spectacles and built form. The very naming of cities is an act of signification. To call a city New Amsterdam is to situate it in a universe of meaning and imagery. And to change it from New Amsterdam to New York is more than just a new entry in a gazetteer. The naming of cities, the mapping of cities, the written and spoken descriptions of cities all constitute forms of urban representation.

Regimes of representation are discourses of meaning which include whole sets of ideas, words, concepts and practices. Regimes are the more general context in which particular forms of urban representation take on specific meaning. Let us illustrate with an example. In a fascinating study Thomas Klak (1994) compares how Havana in Cuba and Kingston in Jamaica were represented in major US newspapers. Both cities face problems of poverty. However, while Kingston was shown to have problems typical of Caribbean islands, the problems of Havana were represented as failure of the communist system. The two cities with similar problems were represented very differently. They were part of different regimes of representation which gave very different meaning and contrasting significance to similar issues.

We use the term regime in much the same way that Thomas Kuhn used the term *paradigm*, Louis Althusser employed the concept of *prob-*

lematic and Michel Foucault defined *discourse*. We prefer the term regime because of its coercive connotations; regime in contrast to *discourse* or *problematic* highlights the social control/contestation implication and foregrounds the issue of knowledge and power. Regimes of representation are politically contested and produced.

We can identify different types of regime. At a very broad level *regimes of representation* are closely connected but cannot be entirely reduced to broader economic and social practices. For example, the development of capitalism involved the creation of new systems of meaning for cities. The religious signification of cities, which had dominated pre-capitalist societies, was no longer paramount. The meaning of cities became incorporated into a commercial system of signification and evaluation. Capitalism is not just an economic arrangement, it was and is a whole new set of meanings.

Regimes of representation are rarely stable. At times of intense instability or in periods of competing regimes, crises of representation can be identified. These occur when the systems of meaning are changing so rapidly and so markedly that there is an intensity of re-representation. In recent years there has been a crisis in urban representation brought about by changes in the spatial organization and reorganization of capitalism at the global level (Lash and Urry, 1994). *The restructuring of the global space-economy is reflected and embodied in the crisis of urban representation.*

There are also what we may term *campaign regimes*. Ward (1996) has described five dominant campaigns: selling the frontier and early town promotion; selling the resort; selling the suburb; selling the industrial city; and selling the post-industrial city. Each 'selling' has a common repertoire of themes and images that emerges from selling the same basic commodity, sharing the same messages and using similar iconography, albeit tailored differently to specific places. Later, we will look at four contemporary campaign regimes emerging from the current crisis of urban representation.

THE GLOBAL REGIME OF URBAN REPRESENTATION

Throughout much of the nineteenth century and most of the twentieth there was a crude division of labor; manufacturing production was undertaken in the industrial cities of the capitalist core economies, while command functions were concentrated in large cities, especially world cities. In the past 30 years there have been major changes in manufacturing production. Technological developments, leading to the deskilling of labor and the decreasing size of transport costs, have allowed manufacturing production to be undertaken around the world. Location is now driven more by labor costs than by the need to be close to natural resources or markets. The net effect is the relocation of manufacturing, a global shift that has seen the decline of older manufacturing cities in the capitalist core economies and the growth of cities in the newly industrializing countries (Dicken, 1998).

New footloose industries have also arisen, especially in the high-tech sectors, which have very different locational requirements from the older metal-bashing industries. They are more concerned with access to information than closeness to a coalfield or sources of electric power. These brain-driven, knowledge industries have a high degree of locational flexibility.

Service and command functions, because of technological changes such as email, video-conferencing, cheaper telephone and fax rates, can also be located away from the previously dominant cities. While there is still the pull of personal contact and inertia, the push of rising costs has allowed the relocation of functions previously tied to the very large cities.

In total, there is a greater locational flexibility in the contemporary economy. This has been reinforced by the growing pool of mobile capital, as witnessed in the growing number of research and development centers, tourists, conventions and rotating spectacles such as the Olympic Games. There is a growing pool of money that can be attracted, with the right mix of incentives and attributes, to particular cities.

These profound changes have led to a new urban order where jobs and investment move quickly and often across the world, from city to city, up and down the urban hierarchy. In this chaotic geography cities need to position and reposition themselves. There is a crisis of urban representation as old images are cast aside and new images are presented for the new urban order.

LIGHT AND SHADOW

Urban representation has two distinct discourses. The first is the positive portrayal of a city; the city is presented in a flattering light to attract investors, promote 'development' and influence local politics. But every bright light casts a shadow. The second discourse involves the identification of the shadow, the dark side that has to be contained, controlled or ignored. This discourse works through silence, as some issues and groups are never mentioned, and through negative imagery, as some groups and issues are presented as dangerous, beyond the confines of civil debate (see Murray, 1995; Wetherall and Potter, 1992; Wilson, 1989, 1995).

The official web sites of cities across the world are good examples of the bright-sided portrayals. A large number of cities have constructed their web sites to provide information on investment, tourism, affairs, education and community. The city of Detroit web site, for example, contains, among others, the facts that the city has hosted dramatically increasing inward investment and that it also has diverse cultural resources (http://www.ci.detroit.mi.us). The city's slogans represent an economically promising and culturally rich Detroit: 'Making it better for you', 'New city for a new century', 'It's a great time in Detroit' and '1998 best sports city'. The opposite representation of Detroit is founded in numerous photo images of America's poorest in its inner city (Bukowczyk and Aikenhead, 1989). Neill's (1995a, b) disturbing metaphor, 'lipstick on the Gorilla', critically insinuates Detroit's failure of image-driven development projects. The promising web sites and the derelict photo images show two different Detroits.

We will discuss three elements of this second discourse, the shadow, that arise from the place wars. The first is the apportionment of blame. In his discussion of the reaction to so-called urban decline in the US, Robert Beauregard (1993) notes that from the early 1960s to the mid-1970s race was a recurring motif in the discourse of decline. The background, of course, was the so-called race riots that rocked US cities in the 1960s. In 1969 Richard M. Nixon noted, 'As we look at America, we see cities enveloped in smoke and flame.' Race was now the cement that held the debate together, a debate that looked at causes and solutions in urban renewal, desegregation, and police methods. But at the heart of the perceived urban problem was a large black population concentrated in ghettoes. The rhetoric of the discourse involved such apocalyptic phrases as 'urban crisis', 'city as wilderness', 'ailing cities' and 'urban society on the verge of collapse'. The Secretary of Housing and Urban Development, George Romney, speaking in 1972, noted that 'the whole social web that makes living possible is breaking down into a veritable jungle.' Even the Democrat senator Robert Kennedy noted, 'We confront an urban wilderness more formidable and resistant and in some ways more frightening than the wilderness faced by the Pilgrims or the pioneers.' Cities were less the promised land and more the American nightmare. And the id of the urban imagination was the black presence. By the 1990s the vocabulary had changed; the terms underclass, ghetto, welfare and a host of others were used, sometimes deeply coded, often not. The problem was the inner-city blacks who were perceived as a cause of crime, urban decay, moral collapse and the decline of the city.

A second and related theme is the issue of social control. Business elites have long had a distrust of participatory democracy. The urban booster rhetoric is no exception. However, in 'democratic' societies the emphasis is not so much on explicitly excluding the 'public' as on encouraging private–public partnerships. This is the buzzword for urban development schemes. In effect, these partnerships disenfranchise the general urban population and allow debates to go on behind closed doors in a cosy corporate compromise. The marginal groups remain marginal and those at the center continue to define the terms, set priorities and allocate both resources to ensure success and blame to account for failure.

A third theme is the foreclosing of alternatives. This is often attempted by linguistic usage. Such terms as *the bottom line*, *fiscal realities* and *the new urban realism* are all phrases that seek not so much to identify fiscal constraints as to close off an alternative discourse about other fiscal priorities and different spending choices. The constant use of natural analogies in urban boosterism and especially the bodily analogies of urban development schemes, for example using the image of revitalizing the heart of the city center for reconstructing the downtown, not so much dramatizes the endeavor but presents it as a wise doctor tending a sick child. Who could possibly argue with such an image? Only someone who was unreasonable, and beyond the reach of everyday common sense.

Marketing the city

Although urban marketing has been practiced since the age of colonial expansion (Ward, 1990, 1994; Ward and Gold, 1994), in recent years it has increased in importance and intensity as cities around the world compete in a crowded global market.

Marketing the city has been discussed by various commentators. Two broad approaches can be identified: first, there is a body of work that ties urban marketing to a deeper political economy (Harvey, 1989; Kearns and Philo, 1993; Logan and Molotch, 1987); second, there are studies which focus on practical marketing strategies (Ashworth and Voogd, 1990; Gold and Ward, 1994; Kotler *et al.*, 1993). Although there is some overlap between the two, the former tends to emphasize the recent transformation in urban governance and the involvement of business coalition in local economic development, while the latter focuses on the detailed processes of place promotion.

In one of the most sustained analyses of city marketing, John T. Bailey (1989) suggested a three-stage evolution. His model was developed for the US although it is suitable for wider generalization, paralleling as it does historical trends in the capitalist economy. The first generation, *smokestack chasing*, was concerned with generating manufacturing jobs through attracting companies with subsidies and the promise of low operating costs and higher profits from existing or alternative sites. The poaching of factories from other cities was, and in some (especially Southern) states still is, a major element of local job promotion. Urban representation centers on low operating costs and availability of subsidies.

The second generation, *target marketing*, involves the attraction of manufacturing and service jobs in target industries currently enjoying profitable growth. There are still attempts at luring plants from other locations, but the promotion also includes improving the physical infrastructure, vocational training and stressing good public–private cooperation. Representation continues to mention low operating costs but includes the suitability of the local community for target industries and the more general notion of good quality of life. Singapore, for example, seeks to attract foreign companies which are associated with high-tech, high value-added manufacturing industries (Economic Development Board, Singapore, http://www.sedb.com).

The third generation, *product development*, contains the objectives of the first two stages but includes emphasis on the 'jobs of the future' while representation now includes global competitiveness, human and intellectual resources as well as low operating costs and quality of life. With each successive stage the message becomes more sophisticated and urban representation has to include issues of quality of life.

Reimagining the city

Haider (1992) uses the term *place wars* to describe the fierce competition between cities for both fixed investment and circulating capital (conventions, spectacles, tourists, etc.). In this competition, emergent world cities no longer have a monopoly of command and control functions (such as banking); industrial cities in the developed world have to compete with places around the world; and all cities compete for the benefits of the postindustrial economy. The global shift of manufacturing and the changing space economy

Figure 7.1 The Everson: an I. M. Pei-designed art museum in Syracuse, NY. (*Photograph by John Rennie Short*)

of a deregulated postindustrial space economy have brought about growing place wars. In these wars, at least four interlocking campaign regimes can be identified:

- World cities and wannabe world cities
- Look, no more factories!
- The city for business
- Capitalizing culture

WORLD CITIES AND WANNABE WORLD CITIES

In recent years London and New York have been slipping in their proportion of world banking, financial markets and corporate headquarters, while the relative position of Tokyo has been increasing dramatically. In the case of New York a series of campaigns have been mounted to secure the city's position as the country's world city. From the hugely successful, earlier slogan of *I Love New York* to the more recent 'The business city that never sleeps' campaign, the city has sought to promote a positive image. While London lacked a similar formal advertising campaign, the iconography of London Docklands was partially aimed at

creating a world city feel, the tall skyscrapers and large groundscrapers all clad in a postmodern architecture all bespoke a city where serious money could be made, monitored, traded and measured. The postmodern referencing of recent building developments in and around the City of London was aimed at creating a more upbeat contemporary business image (Crilley, 1993). All three world cities have been facing competition from what I will refer to as wannabe world cities. These are cities that have some command functions but want more. They include Los Angeles, Atlanta, Chicago, Paris and Birmingham in the UK. It is on this level of the world urban hierarchy that extensive campaigns have been mounted, as one slogan for Atlanta optimistically noted, *Claiming its International Destiny*. In some cases there is a conscious attempt to attract formerly big-city functions like banking. Charlotte, North Carolina, is now one of the largest banking centers in the US. In other cases, there is an attempt to usurp national urban dominance. In Australia, for example, Sydney and Melbourne have long competed to be Australia's world city, the main connection point with the global economy command structure. The

Figure 7.2 The Chicago skyline is a text of signature architects. (*Photograph by John Rennie Short*)

battle was won by Sydney, a victory reflected in the fact that the Sydney Opera House is a recognizable icon in the international community while Melbourne struggles to achieve a distinctive representation.

The wannabe world cities compete for command functions and world spectacles. A good, though not infallible, guide to wannabes is to note those cities that have either hosted or applied to host the Summer Olympic Games (see Table 6.3). Hosting the Olympics gives a city international recognition, a psychological boost and a great opportunity for business, real estate development and increased investment. That is why the competition over hosting the Summer Olympic Games is increasingly fierce. Host and candidate cities between 1992 and 2004 are all to an extent wannabe world cities. Regional or

national centers in Europe, including Amsterdam, Athens, Barcelona, Birmingham, Manchester, Rome and Stockholm, have been seeking for promotion to world centers. Some post-Socialist cities, such as Beijing, Belgrade, Berlin and Warsaw, have been trying to use the Olympics for attracting foreign investment. Istanbul, Buenos Aires, Cape Town and Rio de Janeiro are good examples revealing the efforts and desires of under-represented regions to participate in the global competition.

Wannabe world cities are concerned with ensuring the most effective international image. It is essential to have all the attributes of a world city; these include an international airport, signature buildings of big name architects (e.g. Michael Graves, Arata Isozaki, Philip Johnson, I. M. Pei – see Figure 7.1, John Portman, Richard

Rogers, Aldo Rossi, James Stirling), impressive buildings (the most recently constructed tallest buildings in the world are in Kuala Lumpur, Shanghai, Hong Kong and Taipei) and cultural complexes such as art galleries and symphony halls. Combining these elements is a useful strategy, as in hiring a famous architect to design a cultural complex like James Stirling's Art Gallery in Stuttgart, Isozaki's Museum of Contemporary Art in Los Angeles, Pei's glass pyramid in front of the Louvre in Paris or, one of the oldest yet still impressive, Joern Utzon's Opera House in Sydney. The construction of these complexes includes also urban redevelopment schemes in which local developers, landowners and politicians can make substantial fortunes. Wannabe cities are cities of spectacle, cities of intense urban redevelopment, and cities of a powerful growth rhetoric. These symbolic and concrete imaginations are possible because there are mobile world spectacles and footloose functions, especially financial services, previously fixed in only a few world cities.

Wannabe world cities have an edgy insecurity about their role and position in the world which gives tremendous power and energy to their cultural boosterism. The desperate scramble for big-name architects (see Figure 7.2), art galleries and cultural events is a fascinating part of the place wars in the US among cities aiming for the top of the urban hierarchy.

The struggle is never ending as events can have a negative impact on even the most carefully constructed image. Witness the case of Los Angeles. After several years of bad press including earthquakes, riots and uprisings, mudflows, fires, and scenes of police brutality that resounded around the world, in 1995 the city established the public–private Los Angeles Marketing Partnership (LAMP) which subsequently launched a five-year media blitz aimed at presenting a more positive image of the city, to counter the popular conception of the city as a disutopia sliding into anarchy and destruction. In 1995 $4.5 million was spent on billboards and advertisements initially targeted at the Greater Los Angeles district in order to raise local spirits, generate civic pride and dominate the discourse of the city. The emphasis of the campaign was on the city as a place to do busi-

ness, with such slogans as *Los Angeles is the No. 1 port of entry in the country*, *It exceeds New York*, and *L.A. has a bigger economy than South Korea, and South Korea is booming*. The advertising firm hired to oversee the campaign was Davis Ball and Colobatto whose previous successful campaigns included the rejuvenation of the sales of eggs despite their reputation for high cholesterol. One of their more popular ads featured eggs marching out of a prison with the line *Eggs, give them a break*. Even with the experience of such successful turnarounds under their belt, one of the partners conceded that Los Angeles was a 'very tough sell' (Rainey, 1995). Selling Los Angeles, like selling eggs, was an attempt to polish a tarnished image.

LOOK, NO MORE FACTORIES!

Industrial cities in the developed world have a difficult time in an era of world competition and the global shift of industry towards much lower cost centers. To be seen as industrial is to be associated with the old, the polluted, the distressed and the outdated. A persistent strand of urban (re)presentations has been the reconstruction of the image of the industrial city. The process has been described for a range of cities in a variety of ways:

- reconstructing the image of the industrial city (Short *et al.*, 1993)
- revisioning place (Holcomb, 1993)
- city make-overs (Holcomb, 1994)
- selling the industrial town (Barke and Harrop, 1994)
- gilding the smokestacks (Watson, 1991).

Cities like Manchester in Britain, Syracuse, Pittsburgh and Milwaukee in the US and Wollongong in Australia have all been (re)presented in a more attractive package that emphasizes the new rather than the old, the fashionable postmodern rather than the merely modern, the postindustrial rather than industrial, consumption rather than production, spectacle and fun rather than pollution and work. Emblematic of these shifts in meaning was the change in the logo of Syracuse. This city of over half a million people in New York State

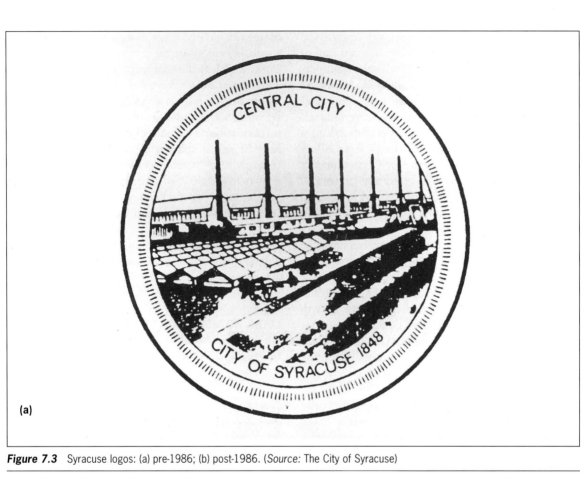

(a)

Figure 7.3 Syracuse logos: (a) pre-1986; (b) post-1986. (*Source:* The City of Syracuse)

(b)

© 1986 City of Syracuse NY

had an industrial history originally based in salt production and later in a range of manufacturing and metal-based production. One side-effect was the pollution of the local lake. It was a typical frostbelt industrial city with an official seal (Figure 7.3(a)) which celebrated its industrial base with images of factories and salt fields. In 1972 the Mayor of the city organized a design competition to replace the 100-year-old seal. There was community resistance and it was only in 1986 that another Mayor was able to introduce a new city logo (Figure 7.3(b)). This logo represents a clean lake and an urban skyline with not a factory chimney to be seen (see Short *et al.*, 1993).

Reimagining the industrial city involves the physical reconstruction of the city. The process of deindustrialization sometimes allows opportunities for urban redevelopment as factories are abandoned and new geographies of production and circulation leave old docks and railway lines economically redundant. The urban redevelopments are schemes to make money but are often portrayed as attempts at reconstructing/reimagining the city in the form of postmodern architecture and postindustrial economies. Baltimore's Harbor Place and Pittsburgh's downtown developments are good examples.

The representations of an industrial city also involve an internal debate as well as the manipulation of external images. Industrial cities have a culture, an emphasis on manual work, a collective sense of meaning and significance tied to their industrial and manufacturing base. Industry not only provides a means of living; it also creates a context for individual and collective identities. Deindustrialization destroys this meaning and representations of a postindustrial city challenge and undermine these identities. Image make-overs are also struggles over which image will dominate. Part of this internal debate involves a renegotiation with the physical environment. Old-fashioned industry was very polluting, and waste was dumped on the ground, in rivers and lakes. There was a toxic legacy. Moreover, there was an attitude and orientation to the physical environment that stressed work. The environment was merely a backdrop, a context and a refuse bin for industry. Reimagining the industrial city involves restructuring the social–environment relationship. In the case of Syracuse the lake was represented in the new logo as pristine and suitable for recreation. The reality was very different as the lake still held toxic waste. There are success stories. In Barcelona the urban renewal associated with the Olympic Games involved the opening up of the old docks to a harbor waterfront with a pedestrianized walkway. Until the mid-1980s, the city had turned its back on the sea, leaving only warehouses and docks at the water's edge. Deindustrialization, changes in transportation and the need for a new image all meant that the harbor front became a place for leisure consumption rather than production and storage.

The internal debate also involves the creation of new agendas and the suppression of old ones. In 1988 John Norquist was elected Mayor of Milwaukee. The old image of the city was as the beer capital of the nation, an industrial city and a place with a rich socialist tradition. Deindustrialization in the 1970s and 1980s had destroyed the old job base of the city, challenged its collective identity and left a vacuum for new representations. Mayor Norquist led an aggressive campaign to (re)present and restructure the city. He trashed the city's welfare tradition and created a pro-business climate by offering competitive (i.e. low) utility rates and tax rates, initiating a series of lakefront festivals and promoting public–private downtown construction schemes including a new convention center. Norquist became a symbol of urban reimaginings and he was widely quoted and feted in the media as 'one of America's boldest mayors' (Eggers, 1993), 'the brave marketeer of Milwaukee' (*The Economist*, 1995), and 'a disciple of the new urbanism' (Auer, 1995). Norquist and Milwaukee have become an exemplar to many industrial cities in the US.

In the current economic climate many city administrations have to present a pro-business image. However, they also have to get (re)elected. While business has the finance, the people have the votes. The reimaginings of the city thus combine a curious mix of pro-business sentiments with shadings of wider social concerns which reflect and embody local political culture. Thus

Glasgow's *City Vision*, launched in January 1996, contains passing reference to quality of life, maximum employment and a caring social infrastructure. This is in rich contrast to the free-market vision enunciated by Mayor Norquist of Milwaukee. The varying forms of urban reimaginings embody differences in community identity, local political culture and the rhetorics that are most likely to carry the day in particular cities in various countries.

THE CITY FOR BUSINESS

The pro-business message has become a standard theme in the reimagining and selling of cities. The hypermobility of capital, the intense and growing competition between cities for both fixed capital investment and a piece of the circulating capital of tourists, conventions and global and national spectacles have all reinforced the age-old basic booster message, 'Do your business in our city' and 'come visit us'. Differences, however, can be noted. Glasgow's *City Vision* launched by the City Council encourages maximum employment as much as technological progress. The *Vision* is promoted by local politicians who have to get elected in a left-wing city and thus need to embody the local political culture while also promoting the city to the wider business community. In the US, in contrast, this left-wing tradition is either lacking, or in the case of Milwaukee, silenced. Moreover, urban promotion in the US is most often a private sector activity, the local Chambers of Commerce being particularly active, either on their own or as the dominant partner in joint public–private initiatives. Thus urban representation in the US more totally reflects the needs of business.

In the section entitled 'Light and Shadow', earlier in this chapter, we mentioned the City of Detroit's web site which shows the city's countless projects assisting local firms from small businesses to large corporations in pursuit of building 'a world class city' and 'a new city for a new century'. The web site of Singapore's Economic Development Board is entitled 'Global city with total business capabilities' (http://www.singapore-inc.com/edb.html). In order to ensure Singapore's competitiveness in the global market, despite the disastrous downturn of the Asian economy, the Board takes a proactive approach to the capability and manpower development. Singapore even operates a program named the International Manpower Programme, helping local companies' worldwide recruitment of professionals. Singapore's global business capability will be furthered, indeed it has been the number one in the world, because the city's initiatives and incentives are continuously reviewed to sustain the pace of investments in manufacturing, services and high-tech industries.

CAPITALIZING CULTURE

Sinclair Lewis's quote at the beginning of this chapter from over 70 years ago still resonates. 'Theaters and art galleries' are still used to attract thousands of visitors, although the attractions have widened to include sports stadia, waterfront redevelopments and historic neighborhoods. While the attractions have changed, the concern with *capitalizing culture* has remained. This campaign regime is found in wannabe world cities, old industrial cities and cities eager to attract the command functions of the world economy and in pursuit of Bailey's third stage of product development.

Art exhibitions and galleries, opera halls, museums, festivals and symphony halls are a vital part in the reimagining of cities. They intimate world city status, a city which can attract and retain the executive classes and skilled workers of the high-tech industries of the present and the future. These cultural attributes are also a source of revenue in their own right. Culture is now big business. Consider the following example. In 1995 the Art Institute of Chicago held a very successful exhibition of Claude Monet's paintings. The exhibition ran from 22 July to 26 November and generated $389 million in economic benefit to the city's economy and a $5 million profit for the Institute (Howlett, 1995). Cultural strategies like these can be defined as attempts to identify, mobilize, market and commodify a city's cultural assets, and are now a major element in urban regeneration and the stimulation of a city's economy (see Bassett, 1993; Griffiths, 1993; Zukin, 1995). One of the most important recent examples was the case of Glasgow and its *Glasgow's Miles Better* Campaign, the 1988 National Garden Festival and its designation as European City of Culture in 1990 (see Boyle and Hughes, 1991). In 1999 Glasgow will be the UK City of Architecture and Design.

Big-name architects designing art galleries is a killer combination for cities. In 1998 the new Getty Center opened in Brentwood, Los Angeles. It had been a long job for the architect Richard Meier, who had worked over 10 years on a project that cost one billion dollars. The new center for the Getty collection sits atop a hill at the intersection of highways 10 and 405. It marks what one commentator noted as a 'hunger for respect, for class' (Andersen, 1997: 66). The Director of the Center was quoted as saying 'I think it will make it easier for serious people to persuade themselves they might come and live in Los Angeles. It will bring tourists here, and that will change the caricatured view of Los Angeles' (*ibid*.: 72). Another architect, Frank Gehry, who had failed to make the shortlist for the Getty Center, was the designer of the Guggenheim Museum in Bilbao, Spain. The city is the capital of the Basque region, and the Museum marks the prestige of the city in the world. The project began as a piece of urban and cultural boosterism. The Basque regional government agreed to pay for the project and the Guggenheim Foundation lent its name and its collection in return for a fee and control of the project. Three architects were asked to submit designs: Arata Isozaki, Coop Himmelblau (a firm in Vienna) and Gehry. His final design is a voluptuous building with curving titanium panels.

Urban cultural capital includes more than just traditional elements of so-called high culture. Popular culture in a variety of guises is also important. There is the culture of leisure; cities now represent themselves as fun places, places where the good life is attainable. The good life is increasingly defined as not only lucrative employment but also ample time for leisure. The leisure industry is a major growth area and source of revenue and economic development. Resort promotion has long been a feature of place promotion. But now the cities where work is done also advertise themselves as cities where leisure is pursued. The marketing of the city as a center for play has been tied to dining, shopping, nightclubbing and outdoor pursuits. Four themes in particular can be noted. *The Historic Feel* plays up the historic connections of the city or particular city neighborhoods, while sometimes whole districts are historicized, including 'antique signs', 'authentic' landmark sites and a guided historical narrative in the form of routes, maps and brochures. Robert Hewison (1987) has referred to the heritage industry in Britain while Boyer (1992) writes about merchandising history. Cities are being historicized as part of urban make-overs, as revenue-generating schemes as part of the attempt to give a positive image of the city by highlighting its historical depth. *The Festival Package* emphasizes frivolity, resorts and spas, sporting events, shopping malls and convention centers. Mexico City, Miami, Milwaukee and New Orleans, among many others, use this package to sell themselves in the popular press. The Indianapolis 500, the greatest spectacle in auto racing, has been attracting a large number of fans around the world annually, and its TV coverage has promoted the vibrant image of the city with a huge motor speedway. *The Green and Clean Theme* situates the city in a postindustrial world with clean air, good beaches, easy access to the 'natural' world, and active recreation facilities such as sailing, fishing, swimming and biking. San Francisco, San Diego, Seattle, Portland, Vancouver and Sydney work this theme into their urban representation. *The Package of Pluralism* highlights the rich ethnic mix that leads to a varied urban experience, including specialized shopping centers, ethnic restaurants and ethnic carnivals. New York, Toronto, Los Angeles, Chicago and San Francisco sell this theme.

These four themes are not exclusive and individual cities may use elements of all of them in their representation.

Popular culture also includes spectacles, festivals and sports. Professional sports teams are increasingly important in the representation of US cities, highlighting a number of more general ideas about capitalizing culture. Professional sports in the US, particularly basketball (National Basketball Association), football

(National Football League), baseball (Major League Baseball) and of slightly lesser importance ice hockey (National Hockey League), are big business. According to Kotler *et al.* (1993: 44), a professional sports team is a powerful image-generating machine and an economic anchor for a city. Without the Packers, who in the US outside Wisconsin would know of the small city of Green Bay? Molotch (1976: 315) points to the importance of a professional sports team carrying the city's name. The athletic teams are, he notes, 'an extraordinary mechanism for instilling a spirit of civic jingoism regarding the progress of the locality'. Many cities have made considerable efforts to build a new sports stadium with the intent of attracting a professional sports team that would provide them with big-city status, a vital and youthful image, and a powerful vehicle for economic development (Wulf, 1995; Zelenko, 1992). Kotler *et al.* (1993: 39) call these efforts the 'stadium-mania' of American cities.

The sports franchises are located in cities around the country. They make money for their owners and the city revenues but also give recognition to the cities. The New York Yankees, the LA Clippers, the Tampa Bay Buccaneers and the Philadelphia Flyers are all examples of professional sports teams with an accompanying city name and associated image. This is the rule rather than the exception. Cities want the franchises for the money they directly generate and for the image they represent. To have an NBA, NFL, MLB or NHL team is to play in the big leagues, to have your city's name mentioned in the extensive sports coverage that saturates the mass media. A successful sports team is great public relations for the city. One city even built its reimagining around professional sports. Indianapolis initiated a new public relations exercise and economic development strategy in 1982 aimed at coordinating development strategies and improving the image of the city. Over $1 million was spent over four years. As a central part of the representation of the city, the NFL Colts were lured from Baltimore in 1983 to begin the 1984 season as the Indianapolis Colts. Professional sports teams provide revenue, taxes, a possible source of civic pride and name recognition in the national and world media.

The slogan of Jacksonville, *The expansion city of Florida's first coast*, is more meaningful when you realize that the city was recently awarded an NFL expansion team. Having obtained an NFL team, the city used the fact in its promotional advertising.

The very attractiveness of sports franchises gives the owners leverage as cities compete one with another in order to keep their teams, lure existing franchises or snag expansion teams. In recent years owners have moved their teams in order to get better stadiums with higher revenue-generating facilities, sweetheart deals and tax breaks. In the NFL alone in the last decade the Cardinals have moved from St Louis to Arizona (1988), the Rams have moved from Los Angeles to St Louis (1995), the Raiders have moved from Los Angeles back to Oakland (1995), and the Browns were scheduled to move from their long-term home in Cleveland to Baltimore which had lost the Colts to Indianapolis in 1984. The same story can be told for the other big sports. Despite the relocation, some teams retain the former names with only the city name changed. Thus the name of the Utah Jazz from Salt Lake City is understandable only when you know that this NBA team relocated from New Orleans in 1980, and Los Angeles Lakers originally were the Minneapolis Lakers. Crothers (1995) writes of the shakedown as owners move or threaten to move if new stadiums are not built, and if better leasing deals and larger tax breaks are not forthcoming. He lists three NHL teams, eight NBA teams, 14 NFL and 14 MLB teams negotiating for better arrangements with the threat of moving if their demands are not met. Culture can be capitalized, but do not be surprised if your cultural capital threatens to move to another city.

Urban representation and urban boosterism

The representation of the city is becoming closely associated with the marketing of the city. Urban representation and urban boosterism now go hand in hand. This is becoming evident around the world but is most marked in the US. We can look at the US as an exemplary case of the close connection between urban representation and urban boosterism.

In the United States there has been a much more fluid regime of representation. The sheer size of the country, its lack of urban primacy and its recent urban competition have all militated against very fixed urban hierarchies or systems of representation. In comparison with the more centralized control of local authorities throughout much of Europe, the more decentralized governmental system in the US also allowed local governments much more opportunity for place promotion and urban boosterism. While local authorities in Britain, for example, had severe restraints on how much they could spend on place advertising, no such limits applied to cities and states in the US. There has also been a constant need for place promotion in the USA. In the earlier years cities had to be sold to foreign investors and migrants from overseas.

The eighteenth and nineteenth centuries are full of examples of cities representing themselves as sites of profitable investment and suitable destination points for ambitious and hardworking immigrants. In the rolling western frontier, cities had to sell themselves as part of land development opportunities associated with the coming of the railway. After the Civil War urban boosterism in the South targeted northern investors. Atlanta, Georgia, has had one of the most sustained urban booster campaigns of the twentieth century, stretching from the *Forward Atlanta* campaign aimed at northern industrialists at the turn of the century, through the lobbying for an international airport in the 1940s to the hosting of the Olympic Games in 1996.

William Cronon (1991), in his book *Nature's Metropolis*, describes the opportunity for urban growth in the Midwest in the middle of the nineteenth century and the various attempts by cities to represent themselves as *the* city of the region. The whole region was pregnant with urban opportunity before Chicago emerged as the dominant city. Urban representation and urban boosterism went hand in hand in the US. The very names conjured up images of mythical renewal (Phoenix) and classical splendor (Syracuse). The advertising campaigns played on the city as a place to do business. The growth machine defined the civic good and was the dominant force in representing the city (Logan and Molotch, 1987). It still is. In comparison with Britain, where place promotion was a responsibility of local government, place promotion in the US has always been dominated by the business community through such organizations as the Chambers of Commerce, Economic Development Bureaux and *ad hoc* arrangements, sometimes in association with the municipalities but often not. In the fluidity of the US urban system it became paramount for cities to have a strong and positive image. The images revolved around the central concept of the city as a place receptive to mobile capital.

A variety of strategies have been discussed in this chapter: attempts at world city status; the distancing of cities from their industrial past; the presentation of cities as attractive places for footloose business and investment; the city as a place of culture. These (re)presentations are for both external presentation and internal consumption. There is, in effect, a struggle for the meaning of cities. All discourses have their silences. In the new representations, more is said about the city as a place for business, for work, attractive to the senior executives and the governing class of the business community, and much less is said about the city as a place of democratic participation, as a place of social justice, as a place where all citizens can lead dignified and creative lives.

Representations of the city are not politically neutral, neither are they devoid of social implications. The dominant representations play down equality, social justice and an inclusive definition of the good city. There is need for an alternative representation, for urban imagineers who can represent the just, fair city. In the old radical rhetoric, who owned the means of production was a major question. It still is. But who owns the means of reproduction is also important. These are controlled or dominated by the urban boosters and the business community. Indeed the place wars have reinforced the business boosters. In such a competitive atmosphere it is all too easy to succumb to the dominant rhetoric. But the attempt should be made if we are not to lose the image of the city as a just, fair and decent place to live for all its citizens.

PART FOUR

Political Globalization and the City

8

Political Globalization

The thinking underlying this approach – it might even be termed the Summers doctrine – has three elements: a faith in the power of global capitalism to improve the livelihood of humankind; an acknowledgement that free markets, nonetheless, occasionally need government intervention to lead them in the right direction; and a belief that only the United States has the power and inclination to provide the leadership on a world scale.

(Cassidy, 1998: 56)

Economic and cultural globalization, coupled by the collapse of the Soviet Bloc and the disintegration of the bipolar world order, has raised the possibility of political globalization which can be defined as the emergence of global governance and a global polity. To speak of global governance and a global polity can be misleading as it conjures images of a world government, a single unitary and centralized state, controlling the whole world. Waters (1995) notes that political globalization may be defined as a series of complicated processes, including reduced state intervention, globalized political issues, the increasing number of international organizations, the formation of regional blocs and the worldwide spread of liberal democracy. We may add the recent increase of immigration which has eroded many nation-states' control over their borders and complicated the simple notion of citizenship, as we have seen in Chapter 6.

In this chapter we examine the three major factors causing 'denationalization' of the contemporary world:

- the proliferation of supranational organizations and their strengthening power

- the rise of market forces against states
- the challenge of immigration on sovereignty and citizenship.

We also pay attention to the opposing trend of 'renationalization'.

Supranational organizations

Held (1995) outlines a series of steps toward global governance (Figure 8.1). The growth of international and transnational organizations and collectivities has eroded the distinctions between external and internal affairs, and between international and domestic policies. Everything is connected in the age of globalization. Global interconnectedness through economic and cultural globalization, has evidently decreased the power and effectiveness of policy instruments used by the state to control activities within and across its borders. The rapid growth, for example, of capital flows crossing borders can threaten inflation measures, exchange rates, taxation levels and stock price among other government policies. Indeed, much of the traditional domain of state activity and responsibility cannot be fulfilled without resort to international forms of collaboration in the contemporary world. Individual states are no longer the only appropriate political unit for either resolving key policy problems or managing a broad range of public functions. States have been encouraged, forced in some cases, to increase their political integration with other states and to make multilateral negotiations, arrangements and institutions to control the destabilizing effects that

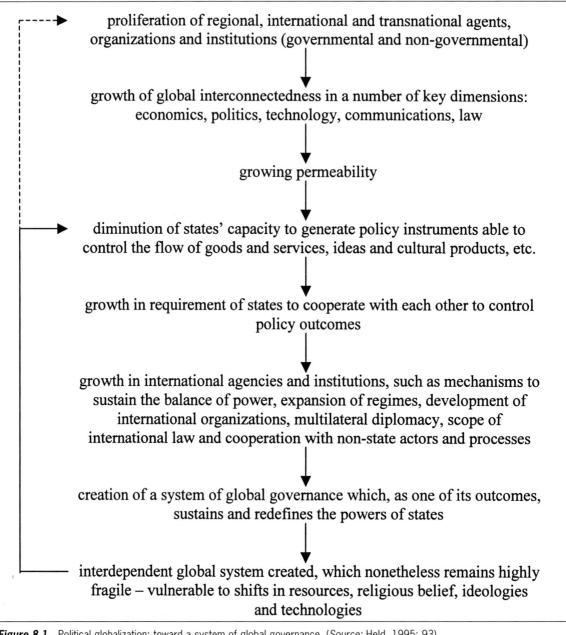

proliferation of regional, international and transnational agents, organizations and institutions (governmental and non-governmental)

↓

growth of global interconnectedness in a number of key dimensions: economics, politics, technology, communications, law

↓

growing permeability

↓

diminution of states' capacity to generate policy instruments able to control the flow of goods and services, ideas and cultural products, etc.

↓

growth in requirement of states to cooperate with each other to control policy outcomes

↓

growth in international agencies and institutions, such as mechanisms to sustain the balance of power, expansion of regimes, development of international organizations, multilateral diplomacy, scope of international law and cooperation with non-state actors and processes

↓

creation of a system of global governance which, as one of its outcomes, sustains and redefines the powers of states

↓

interdependent global system created, which nonetheless remains highly fragile – vulnerable to shifts in resources, religious belief, ideologies and technologies

Figure 8.1 Political globalization: toward a system of global governance. (*Source:* Held, 1995: 93)

can readily spread beyond national boundaries. The result has been a vast growth of institutions, organizations and regimes that have laid a basis for the orderly management of global affairs, that is, global governance.

The proliferation of quasi-supranational, intergovernmental and transnational forces is a very important sign of both the decrease of state power and the increase of international cooperation. The increased number and power of intergovernmental organizations (IGOs), non-governmental organizations (NGOs), international laws and multinational firms have been broadly acknowledged (Weiss and Gordenker,

1996; Diehl, 1997). The growth rates of IGOs and NGOs have far exceeded that of states over the past few decades. By 1996, 38,243 NGOs and 5,885 IGOs had been founded, while the total number of states has never exceeded 200 (Union of International Associations, 1996/1997). A significant number of declarations regarding human rights, securities, child protection and the environment, among other issues, have also been put into effect. Many cooperative international initiatives, such as the Rio Declaration on Environment and Development, have both defined and sought to manage global issues (Laferriere, 1994).

One such example of increasing international cooperation is well presented in an article entitled 'Controlling the global corruption epidemic' (Leiken, 1996–97). The author strongly contends that transnational bribery costs American jobs and, at the same time, costs developing countries efficiency and credibility. In order to eliminate large-scale corruption around the world, he argues the urgent need for more outspoken, better co-ordinated 'US' policy, for deregulation, decentralization and the simplification of government procedures in developing countries and for international instruments against bribery. He maintains that delivering bribes is a global issue because it is against free trade, development and democracy, and thus the 'US' should continue pressing other countries to implement its antibribery recommendations and to select concrete international instruments to facilitate criminalization. According to Eigen (1996), a non-profit organization known as Transparency International (TI) has promoted numerous national chapters combating corruption around the world. In this context, corruption in Nigeria, the most corrupted country in the world (*The Economist*, 1997c), is a problem that the whole world has to be concerned about and one that the international community should continually request the Nigerian government to correct.

The global management of issues is not independent of national interests and national power. The United States has a dominant role in the new world order. It is no accident that both the IMF and World Bank are headquartered in Washington DC, only two blocks away from the White House. The quote at the beginning of this chapter is taken from a profile of Lawrence Summers, the Deputy Secretary to the Treasury (Cassidy, 1998). The Summers doctrine combines both global issues and US national interest.

The worldwide spread of democracy has also intensified the processes of political globalization. Eclipsing any other possibilities of the political system, the promotion of democracy has established a cosmopolitan democratic ideal with global reach (Held, 1995). Countries around the world are increasingly adopting similar policy instruments, policy-making processes and political agendas. Robinson (1996) calls the emergence of a global civil society the twenty-first-century world order.

The market versus the state

An article in *The Economist* (1997e) makes a bold statement, that 'In the global struggle between state and market, markets have gained the upper hand.' The future of the state, the article suggests, will be determined by globalization which prioritizes economic efficiency issues over concerns of social redistribution. There are, of course, heated debates about whether it is true that states are being swept away by market forces. Some define globalization as 'a globally authoritative paradigm' for interpreting and explaining what is happening in the contemporary world (Robertson, 1997). The declining states thesis may be applicable only to certain parts of the globe, leaving out many countries with powerful governments. In this section, however, we review the theory that argues for the enlarging and deepening domain of global market forces and the inevitable decline of nation-states. The counter-thesis arguing for the survival of the state is detailed in the section on the limits of globalization.

The indomitable power of economic globalization reshaping the whole world has been reviewed in Chapter 2. Almost every country across the globe is undergoing some kind of restructuring program in pursuit of a better position in the crowded world market. In order to advance their countries' competitiveness over others, governments are setting up business-friendly environments including deregulation, subsidies and labor supplies. Many developing countries promote foreigner-friendly policies targeted at international footloose capital. The Mexican Investment Board (MIB), armed with the slogan 'Mexico your partner for growth', is advertising the Mexican economy's high ranking in terms of attracting global investment (http://www.mib.org.mx). Singapore, the world's best city for business and the best city in Asia to work and live in, claims credit for its pro-business government, world-class infrastructure, internationalization and excellent science, technology and R&D capabilities (http://www.sedb.com.sg/how/introduction, html).

Growth and competitiveness are being pursued by both the Left and the Right. It is not unusual to see even left-wing parties and governments placing the aggressive promotion of economic growth at the front line of their political agendas. Their electoral promise is to attract foreign direct investment and to promote public–private partnership. In Germany, Gerhard Schroder, the successful Social Democratic candidate for Chancellor in the 1998 national election, admitted the limit of the welfare state and the outdatedness of left-wing discourses. He and his party spoke of building a strong German economy for the next decade, but rarely used traditional socialist rhetoric.

Alongside international (non)governmental organizations, transnational corporations often affect many state policies with relation to tariffs, wages and foreign investment. The pressures from the World Bank and the World Trade Organization (WTO) on developing countries to open their domestic markets have actually been interwoven with multinational firms' interests. The recent financial crisis in Asia has well illustrated the overpowering presence of international organizations, particularly the International Monetary Fund (IMF), the decreasing capability of state governments to manage troubles in domestic economies, and the (in)visible impact of transnational corporations and banks on the bailout negotiations between the IMF and Asian governments, including Indonesia, Korea, Malaysia and Thailand. IMF bailout packages for the ailing Asian economies have strictly requested a significant opening of the financial markets that previously afforded limited access to foreign investors. In the case of Korea, Citibank and Chase Manhattan Bank, which have strongly expressed their interests in merging some insolvent Korean banks, have been (in)directly involved in the negotiation of the IMF and the Korean government (*Dongailbo*, 1998).

Immigration challenges to the nation-state

It has been argued that the rise of migration flows has lessened the capability of the nation-state to control its borders (Joppke, 1998; Sassen, 1996, 1998). Hispanics in California, 'guest workers' across Europe and war refugees throughout Africa have caused a gap between restrictive policy intention and an expanding immigration reality.

In the 1990s, according to Schuck (1998), the US admitted approximately 800,000 foreigners each year for permanent residence, although the number has fluctuated considerably. More than one million aliens enter the US illegally each year and some 250,000 to 300,000 of these remain in illegal status more or less permanently. They have produced a large number of illegal migrants now estimated at greater than five million. Alongside the increasing number of (il)legal immigrants, the resistance of immigrants to assimilate is a profound challenge to the traditional framework designing the nation-state, nationality and citizenship. As we discussed in Chapter 6, many contemporary immigrants maintain multiple citizenships (post-national membership) and multicultural identities.

Sassen (1996, 1998) notes that globalization processes have developed two contradictory trends at the same time: denationalization in the circulation of goods and capital, and renationalization in immigration policy. She concludes, however, that renationalization has its limits, as economic globalization and international human rights norms and discourses are reducing the autonomy of the state in immigration policy.

Many states are resisting the gradual denationalization of citizenship. Along with sovereignty and exclusive territoriality, citizenship is still a critical institution of the modern nation-state. Many states still maintain a policy that forbids dual citizenship. While the numbers of (il)legal immigrants are growing, many countries, particularly developed countries, have been trying to keep the door closed to illegals. Mitchell and Russell (1996) note that Western European governments have made concerted efforts to develop new and tougher forms of immigration control. In the US, the political and social status of illegal immigrants, mainly from Mexico and the Caribbean, has been jeopardized in recent years by the Republican-controlled Congress. There is a growing consensus among states to lift border controls for the flow of capital, information and services and, more broadly, to further globalization. But when it comes to immigration and refugees, whether in North America, Western Europe or Japan, states claim their sovereign rights to control borders.

Following the creating of the single market in Europe, labor migration to and within the European Union has increased (Jenkins and Sofos, 1996). Citizenship of the EU, however, can only be acquired by individuals holding citizenship in one of the member states (Mitchell and Russell, 1996). Koslowski (1998) points to the paradoxical coexistence of two contradictory migration regimes in the EU: an established regime enabling the free movement of EU nationals within the Union states; and an emergent regime restricting the access of non-nationals to Union territory. EU citizenship is entirely dependent on a traditional conception of national citizenship.

Nationality and citizenship are still stubborn facts of political life that resist the easy adoption of global identity.

The enduring nation-state

Despite a general agreement on the increasing erosion of state power by the empowerment of international organizations, the growth of multinational firms and banks and the global spread of democracy, there is solid evidence of the durability of the state system. Although various international initiatives intervene in the processes of domestic policy making, the prevailing territorial sovereignty of nation-states still plays a dominant role in their citizens' political life (Boyer and Drache, 1996; *The Economist*, 1997e; Freeman, 1998; The World Bank, 1997). Held (1995) notes that the impact of the globalization process is highly likely to vary under different international and national conditions – for instance, a country's position in the international division of labor, its place in particular regional blocs, its overall relation to major international organizations. In this section, we discuss the limits of political globalization.

There is still a strong presence of states and central governments. Swyngedouw (1996) argues for the re-scaling of the state in an increasingly globalizing and intensely competitive world economy. Like many other regulation advocates (Jessop, 1994; Peck and Tickell, 1994), Swyngedouw places the state on a more central stage in the understanding of profound social transformations over the past decade. As many supranational and subnational institutions and sources of private capital have been involved in urban/regional restructuring in the era of globalization, nation-states are expected to mediate all sorts of conflicts and tensions between these new institutional forms. Through a case study of closing Belgian mines, Swyngedouw notes that more authoritarian or strong forms of governance, named the 'glocal state', have become increasingly important in order to produce, control and supervise social and physical space necessary for globally induced restructuring.

A diminishing role of the state apparatus is also denounced by a case study of changes in the governance of London. After the dissolution of the Greater London Council in 1986, decision making in the urban policy and planning of London has involved a large number of local governments and *ad hoc* organizations entailing public–private partnership. Newman and Thornley (1997) maintain that this fragmentation of policy making has created a more centralized role for central government to control multiple institutions. The British central government is playing a very important role in promoting London as a world city. The Japanese government is also the principal actor of globalization processes of its capital city, Tokyo (Machimura, 1998).

These case studies argue that specific national, political and institutional contexts must be considered in the debate of the role of the state in today's world. In some places, states are very resilient to the process of globalization. Instead of fading away, they have a new or expanded role to play within a more effective and extensive system of regulation.

The nation-state remains the crucial institution in issues of real income redistribution. The poor get a fairer share in big-government countries relative to small-government ones (*The Economist*, 1997e). Boyer and Drache (1996: 21) argue that 'Efficiency, profitability and competitiveness have not won the hearts and minds of people worldwide.' For many, the danger of globalization is the jettisoning of traditional social welfare policies as national governments aggressively pursue increased competitiveness.

The traditional configuration of the world, divided into almost 200 nation-states as 'bordered power-containers', may not match the prevailing notion of globalization as the explanation of today's world. The world has certainly changed. National boundaries have been eroded dramatically by the global flows of goods, capital, people and information. But we may raise the question: 'Does this mean that states are fading away?' Not really. The supremacy of the nation-state has surely been challenged by market forces, supranational institutions and immigration, but the state continues to be 'the chosen instrument for the organization of society' (Boyer and Drache, 1996).

Nation-states, including their organizations, policies and authorities, have evolved and adapted to new global forces. The state is redefining its role rather than disappearing from the stage. In the process of redefining, repositioning and rescaling of the state there has been the development of an entrepreneurial politics. We will discuss this development in the next chapter.

9

The Entrepreneurial City

In the case of the Manchester Olympic bid, a buc-caneering free spirit Bob Scott and friends occupy a discursive and institutional space opened up through struggles against the backdrop of power-ful globalizing and neoliberalizing tendencies.

(Cochrane *et al.* 1996: 132)

The period from the 1930s through to the 1970s was the high point of social Keynesianism in the Western world. Governments were committed, in varying degrees, to full employment and a bal-anced space economy. The New Deal, the United Kingdom Health Service, the Great Society and the Scandinavian social democracies all provided models for state policies that marked a compro-mise between capital, labor and the state. Since then most Western democracies have seen a retreat from Keynesianism and policies of income redistribution. The main goal of government policy has been the creation of wealth rather than its redistribution. The aims of full employment, eradicating poverty and inequality and the cre-ation of a balanced space economy have been jettisoned or much reduced. Reagonomics and Thatcherism marked a watershed: a definite cutback in welfare spending and an emphasis on effectively responding to market forces. Competitiveness has now replaced fairness as a primary aim of government.

This change, which we can refer to as the shift from the Keynesian State to the Competitive State, has had a marked effect on cities. Channeling money into programs to end urban poverty is now hard to find in the UK, Canada and the US which used to do so up until the 1970s. Cities are much more on their own. Cities are forced to take entre-preneurial approaches to their economic develop-ment.

We should be wary of seeing a homogeneous entrepreneurial city emerging mechanistically from global forces. Hall and Hubbard (1998) have brought together a number of case studies which highlight the differences in the entrepreneurial response. The entrepreneurial city emerges from economic globalization but its shape and nature very much depend upon national regulation sys-tems and local political cultures.

In this chapter we examine the rise of the entrepreneurial city in two ways: the shift of urban governance toward entrepreneurialism and the decoupling of urban from national economies.

The rise of the entrepreneurial city

We can define the entrepreneurial city in a number of ways. Painter (1998) provides a range of mean-ings involved in the notion of the rise of the entrepreneurial city, including the city as setting for entrepreneurial activity, increased entrepre-neurialism among urban residents, a shift from public sector to private sector, and a shift in urban politics and governance from the management of public services towards the promotion of econ-omic competitiveness. It is this last theme that we will examine in this section.

In many cities business elites have dominated the civic discourse even before the onset of rapid economic globalization. The line between civic and business interests was imperceptible in many

US cities. This nexus of city and business interests was formulated into the notion of the city as growth machine by Molotch (1976) and Logan and Molotch (1987). In his 1976 paper Harvey Molotch defined the city as a growth machine since this was the predominant emphasis of locally based elites whose members were involved in local politics. The paper drew upon an emerging literature on urban political economy to paint a picture of city governments dominated by the ethic of growth and the influence of business. Later, Logan and Molotch (1987) extended this argument in a book, *Urban Fortunes,* which identified the growth machine as consisting of politicians, local media, utility companies, universities, organized labor, small retailers and corporate capitalists. This coalition harnessed local politics in the drive to increase land values and create a consensus towards growth and against alternative conceptions of urban organization.

The book generated an enormous amount of interest, comment and criticism. At one and the same time the book was seen as important, enlightening and deeply flawed. We can tell something about the power of an argument by the extent of the critique. The book has come under intense scrutiny. Among the criticisms are that it focuses too much on increase in property values; it is too mechanistic with even the use of the metaphor of the machine giving little opportunity for resistance, contingency, agency and the development of alternative coalitions; and it has a view of politics that prioritizes the formal politics of land-use planning and ignores the world of social reproduction. All good ideas seem to go through a three-phase cycle. In the first they are seen as a great idea, then not so great, until they are considered not great or even much of an idea at all. The work of Logan and Molotch has passed through each of these phases and it is now time for the next phase, which is to use their insights and build upon their work rather than spending too much time repeating what they missed or what they got wrong. All work is provisional and contingent, capable of improvement and elaboration. Logan and Molotch's work was an understanding of US cities at one point in time, but it was also something more: it was a challenge to construct an urban political economy that was theoretically informed and empirically grounded. It is time to go beyond the easy criticisms of the growth machine thesis and take up their challenge.

The connection between the entrepreneurial city and economic globalization has been the subject of much debate. In a celebrated paper David Harvey (1989) argued that the globalization of late capitalism had produced a shift in cities from managerialism, the concern with the allocation of resources, to entrepreneurialism, the encouragement to private capital. In the age of globalization, local politics have gained importance as a focus for proactive development strategies (Mayer, 1995). Cox (1993) wrote of the new urban politics that recombined global economy and local politics in new configurations. In a case study of this new urban politics, Lustiger-Thaler and Shragge (1998) examined the defeat of the left in Montreal in 1994.

Economic globalization has had a marked impact on the way in which cities are governed and the role of municipal governments. It is at the local level that negotiation takes place with capital, in order to insure investment. Montreal is bidding in a competition between cities for investment while other aspects of the city become politically less significant, even depoliticized. To attract investment, the city has to present an image of an efficient economic manager. Machimura (1998) points to the symbolic use of globalization in Tokyo's urban politics which legitimizes the city's emphasis on urban growth.

A substantial number of studies have focused theoretically and empirically on the transformation in urban governance from the welfare-state model towards an economic-development model in the European and North American context. Table 9.1 lists a range of cities studied from the entrepreneurial city perspective. Postindustrial cities, which have gone through the processes from industrial prosperity, to economic decline, and to recent revivals, such as Glasgow, Birmingham, Sheffield and Syracuse, are evidently favorite subjects of entrepreneurial city studies. Olympic (candidate) cities, such as Manchester, have also been examined in terms of the extensive involvement of business coalitions in policy making and planning. This table is also indicative of the bias in the academic 'production' of entre-

Table 9.1 Entrepreneurial cities identified in the literature

Region	Cities (authors)
USA	Columbus (Cox and Jonas, 1993)
	Indianapolis (Rosentraub *et al.*, 1994)
	Los Angeles (Fulton, 1997)
	Miami (Nijman, 1997)
	Minneapolis (Leitner, 1990)
	New Orleans (Miron, 1992)
	New York (Fainstein, 1991)
	San Francisco (DeLeon, 1992; Leitner, 1990)
	Syracuse (Roberts and Schein, 1993; Short *et al.*, 1993)
UK	Birmingham (Hubbard, 1996a, b; Loftman and Nevin, 1998)
	Bristol (Bassett, 1996)
	Cardiff (Imrie *et al.*, 1995)
	Glasgow (Boyle and Hughes, 1991, 1995; Booth and Boyle, 1993; Loftman and Nevin, 1996; Paddison, 1993)
	Liverpool (Parkinson and Bianchini, 1993)
	London (Brownill, 1994; Fainstein, 1991)
	Manchester (Cochrane *et al.*, 1996; Lawless, 1994; Loftman and Nevin, 1996; Peck and Tickell, 1995)
	Sheffield (Lawless, 1994; Loftman and Nevin, 1996; Raco, 1997)
Others	Rotterdam (Hajer, 1993)
	New Castle, New South Wales (McGuirk *et al.*, 1996)

preneurial cities. Certain cities, for example Manchester and Birmingham in the UK, are repeatedly studied in the entrepreneurial city perspective, while the postmodern slant on this shift has been dominated by the city of Los Angeles, the focus of a large number of urban researchers (Dear and Flusty, 1998; Scott and Soja, 1998; Soja, 1989, 1996).

The list of terms popularly used in the literature might be helpful for understanding the transformation:

- city as a growth machine (Molotch, 1976; Logan and Molotch, 1987)
- urban entrepreneurialism (Harvey, 1989; Leitner, 1990)
- the post-Keynesian state (Gaffikin and Warf, 1993)
- urban regimes (Harding, 1994; Stone, 1989)

- city challenge (Lewis, 1994)
- flagship development (Smyth, 1994).

Other terms include urban boosterism, urban corporatism, urban privatism, growth-coalition and public–private partnership. Although the focus and methods of these studies vary, there is widespread agreement that the shift in urban governance results from growing competition between cities for local economic growth. Globalization of markets, production, technology and finance and post-Keynesian urban policy (Thatcherism and Reagonomics) have significantly contributed to the increase of competition between cities. US cities, in particular, have suffered due to the loss of federal funds in the post-federal period.

Almost all city governments have come to promote growth aggressively on a scope unimaginable just a decade ago. Haider (1992) calls this situation 'place wars' in the 1990s. Severe competition to attract and retain business has forced city governments to introduce a range of policy initiatives, such as enterprise zones, urban development corporations, urban subsidies and public–private partnerships (Gaffikin and Warf, 1993). These programs are intended to make the city more attractive to investors.

Many scholars have asked who are the main beneficiaries of this shift. There are competing interpretations. One, most often represented by Peterson (1981), argues that the growth of cities is to the benefit of all residents because any development project has only positive consequences for the city overall. Contrary to this argument, a group of scholars point out that local economic growth does not necessarily promote the public good. Logan and Molotch (1987) identify politically motivated local elites as the main actor and beneficiary of local economic growth. The authors argue that place entrepreneurs, because of their attachment to land ('local dependence' in Cox and Mair, 1988), strongly encourage local growth for their own gains. In their attempts to promote the economic growth of the city, place entrepreneurs organize the growth (business) coalitions that involve and mobilize local governments to intensify land uses for their private gain of many sorts. Peck (1995) critically calls them 'movers and shakers', Lowe (1993) uses the term 'local hero', while Schneider and Teske (1993) refer to them as 'progrowth entrepreneurs'. Their business interests are established as a political phenomenon and subsequently institutionalized (Peck and Tickell, 1995); once organized, they stay organized.

The institutionalization of the business interests of place entrepreneurs has been encouraged by the ideology and policies of post-Keynesian administrations (Lowe, 1993). Their 'growth ethic' – growth is good – is used to eliminate any alternative visions of the purpose of local government or the meaning of community (Logan and Molotch, 1987; Cox and Mair, 1988). The coalition draws on local histories, culture and images to underpin its activities. Peck and Tickell (1995) summarize the contribution of growth coalition, first, to the subordination of welfarist goals to the overriding imperatives of local competitiveness and growth; second, to an acceleration in place-based competition for both public and private investment; and third, to the formation of a new layer of business–political actors at the local level.

City governments' entrepreneurial approach to economic development has been widely criticized for its lack of social goals (Fainstein, 1994; Hall and Hubbard, 1998; Harvey, 1989; Zukin, 1991). Funds for eradicating urban poverty have been severely cut. Political discourses heavily focused on growth leave out the poor and their social rights. City marketing has produced, throughout the US and the UK, for example, convention centers, festival markets and arts centers appealing to the young, professional and affluent (Boyer, 1992). The old, blue-collar and poor tend to be left out.

City government has become less concerned to control or regulate local business and rather more concerned to promote local economic growth. The entrepreneurial city has a pro-growth orientation which dictates taxation and spending policies, the creation of public–private partnerships, the manufacturing of local consent and the creation of new images for the city. In the rest of this section we will examine this last element by looking at recent advertising campaigns for cities in the US.

MARKETING THE CITY

Reflecting on the current processes of economic restructuring and the accompanying rise of the new urban entrepreneurialism, Paddison (1993: 340) notes that 'the concept of the marketing of cities has gained increasing attention as a means of enhancing their competitiveness.' He also identifies a series of different, but related, objectives of marketing the city: raising its competitiveness, attracting inward investment and improving its image. According to Kotler *et al.* (1993), the targets of city marketing are business firms, industrial plants, corporate and divisional headquarters, investment capital, sports teams, tourists, conventioneers and residents. Cities that fail to market themselves successfully face the risk of economic stagnation and decline. Cities have become commodified, regarded as commodities, packaged, advertised and marketed much as any other product in a capitalist society (Goodwin, 1993).

The city marketing process consists of several phases. Haider (1992) suggests five activities: analyzing marketing opportunities, researching and selecting target markets, designing marketing strategies, planning marketing programs, and organizing and implementing the market effort. Ashworth and Voogd (1990) provide a broader context: analysis of market, formulation of goals and strategies, determination of geographical marketing mix, and elaboration and evaluation. Kotler *et al.* (1993) focus more on setting attractive incentives and promoting the city's values and image so that potential users are fully aware of the city's distinctive advantages.

The primary goal of the city marketeer is 'to construct a new image of the city to replace either vague or negative images previously held by current or potential residents, investors and visitors' (Holcomb, 1993: 133). The improved image of the city can be acquired by an energetic marketing campaign as well as by the reality of economic growth. The presence of two sources in the improvement of city image implies a gap between image and reality. Barke and Harrop (1994) note that images may exist independently of the apparent facts of objective reality through image making in place promotion.

Holcomb (1994) compares image (re)making by the city marketeer to cosmetic 'make-overs'. Thus, image marketing is the most frequently employed approach to city marketing (Haider, 1992: 131), particularly for traditionally industrial cities whose economies are either declining or stagnant (Barke and Harrop, 1994; Watson, 1991).

Interestingly there are remarkable similarities in the images projected by cities. Most cities are trying to create and project an image reflecting a vibrant, growing place with accessible location, reconstructed downtown, and sunny business climate. Other common features in the promotional texts are appealing quality of life and cultural promotions such as fairs, festivals and sporting events (Ward, 1994: 58). Burgess (1982) notes the growing significance of quality of life in image making in her survey of the effects of environmental images on the locational decision of executives. Postmodern architecture (Crilley, 1993) and high-

tech industries (Barke and Harrop, 1994; Holcomb, 1993, 1994; Watson, 1991) have also been photographically portrayed in splendid clarity by declining industrial cities to attain a revitalized image (Hubbard, 1996a, b).

The similar images used by different cities are a function of the similar message to be achieved. Moreover, there are still very few firms specializing in place marketing, and the selling of cities is still in its infancy. Common images mentioned above may be termed the 'official image' of the city (Goss, 1993). Focusing on the printed pictorial images of Australia found in published government propaganda, Ryan (1990) investigates which Australian landscapes have been incorporated into the national stereotype, and which have not. In the same token, Sadler (1993) argues that city marketing operations involve the construction or selective tailoring of particular images of the city. Each city selects and authorizes particular favorable images. There is a gap in the literature as well as in our understanding; the transformation in urban governance is discussed under the heading business agenda (Peck and Tickell, 1995) and the politics of local economic development (Cox and Mair, 1988, 1989) but the authorization of a particular image, the politics of image making, is, as yet, rarely studied.

MAGAZINE ADVERTISEMENTS FOR US CITIES

According to Williams (1980) advertising is a magic system reinforcing particular ideologies and a route through which dominant class images can be constructed, reinforced and replicated. Through an intense focus on commercial advertising and its ideological role, Wernick (1991) places advertising within the cultural formation of late capitalism's 'pan-promotionalism'. There are a number of studies that look at place advertising: Burgess and Wood (1988) report the enormous impact of city advertising on the relocation decisions of small firms in the Enterprise Zone of London's Docklands; Fleming and Roth (1991) analyze how advertisers use the images of place to create an appealing context for marketing a product or a service.

Advertising has traditionally been the main component of local economic development strat-

Table 9.2 Major repertories in city advertisements

	Category	Content
Business (economic benefits)	Pro-business political climate (public/private partnership)	• business-assistance programs • sound fiscal policies • industry-specific taxes and incentives
	Ideal workforce	• young, educated, skilled, hardworking labor force • concentrations of educational institutions • sound work ethic • successful labor/management relations
	High technology (advent of 21st century)	• high-tech industry/university partnership • concentration of high-tech industries
	Solid infrastructure	• transportation (highway, international airport) • telecommunications (fiber-optic networks) • local gas and electricity
	Healthy local economy	• economic stability or fast economic growth • upswing trend in job creation • higher proportion of future's industries
	Central location	• central time zone • proximity to large markets (population) • border cities or coast cities
Quality of Life	High quality of life	• affluent natural amenities (beach, lake, fall, mountain, clean air, etc.) • mild weather (sun, warm climate) • health services (world class hospitals) • low cost of living; wide range of housing options • friendliness
	Distinct lifestyle advantages	• high quality of cultural and recreational activities (museum, opera, symphony orchestra, theater, art center, festival, fairs, professional sports teams, golf course) • historical heritage

egies (Haider, 1992; Ward, 1990). There are a variety of advertising packages used by cities: city guides, glossy brochures, fact sheets, web sites and advertisements in journals. We chose to look at magazine advertising (Short and Kim, 1998). Many cities advertise themselves in major business or travel journals to which their target audience has easy access. We collected 34 cities' advertisements from the periodicals *Business Week, Financial World, Forbes, Advertising Age, Fortune, Historic Preservation, National Geographic Traveler* and *New Choices for Retirement Living* for the period 1994–95. Some of these adverts are major texts; the page count for advertising supple-

Table 9.3 Slogans of city advertising

	City	Slogan
Business	Atlanta, GA	Strategically located for global business
	Baltimore, MD	More service, more choices
	Boston, MA	Progress through partnerships; America's walking city
	Chicago, IL	At the heart of everything
	Dallas, TX	The city of choice for business
	Fairfax, VA	The 21st century's first destination
	Jacksonville, FL	The expansion city on Florida's first coast
	Kansas City, MO	America's smart city
	Los Angeles, CA	Capital of the future; Together we're the best
	Memphis, TN	America's distribution center; The new gateway to the world
	Miami, FL	The birth of the new Miami
	Milwaukee, WI	The city that works for your business
	New York, NY	The business city that never sleeps
	Norfolk, VA	Where business is a pleasure
	Oak Ridge, TN	When it comes to technology . . . Oak Ridge means business!
	Philadelphia, PA	The real Philadelphia story; All roads lead to Philadelphia
	Phoenix, AZ	Moving business in the right direction; The best of the best
	Pittsburgh, PA	America's future city: a model of post-industrial economic renaissance
	Rochester, NY	The world's image center
	San Jose, CA	Capital of Silicon Valley
Quality of Life	Atlantic City, NJ	America's favorite playground
	Denver, CO	A cultural and environment adventure
	Hampton, VA	From the sea to the stars
	Lexington, KY	The gateway that's not far away
	Lincoln City, OR	Here, the sights see you too
	Louisville, KY	Your kind of place
	Miami, FL	Perfectly seasoned
	Nashville, TN	Music city
	Omaha, NE	Wild creatures loose in the city
	Orlando, FL	Sun and Fun; You never outgrow it
	Salisbury, NC	Where the past is still present
	San Antonio, TX	Something to remember
	Santa Fe, NM	Where traditions live on
	Saratoga, NY	Discover the magic!
	St Augustine, FL	Your place in history

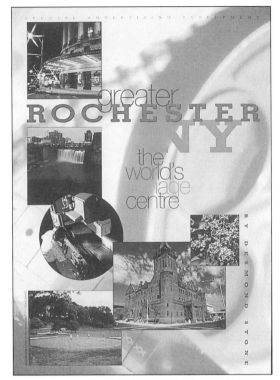

Figure 9.1 City image: Rochester, NY. (*Source: Forbes*)

Figure 9.2 City image: Los Angeles, CA. (*Source: Forbes*)

ments for Boston, Kansas City, Milwaukee, Philadelphia and Rochester regularly exceeded 20 pages. We focused on the slogans and images signified by each city.

We identified a number of recurring themes. Two broad themes can be identified. The first is *the city of work*, and is concerned with attracting fixed capital investments. Here the emphasis was on attracting business by representing the city as a place of low taxes, cheap and docile labor, and in general presenting the image of the city as a pro-business environment. The second theme, *the city at play*, is the attempt to attract the circulating capital of consumers and conventions, the emphasis on what Sinclair Lewis's character referred to as capitalizing culture. The distinction is becoming less clear as the two themes have become more interrelated in recent years.

The subthemes that appear again and again in the ads are outlined in Table 9.2. Given the similar audience of business executives, it is not surprising that the same themes reappear in the

advertisements for very different cities. The similarities arise also from the common use of just a few advertising agencies. A number of agencies specialize in marketing places; Leslie Singer Design, for example, designs most of the city advertisements in *Forbes*. Most ads mention the presence of internationally recognized companies, which give credibility to the discourse of business success and confidence, sometimes reinforced by quotes from CEOs and senior executives. Independent surveys which give the city a high

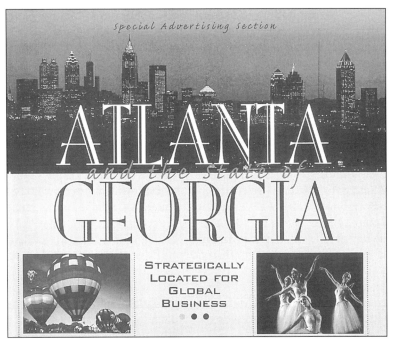

Figure 9.3 City image: Atlanta, GA. (*Source: Forbes*)

rank in terms of quality of life as well as business indicators are also used, while the photo-imagery captures scenes of nature, postmodern architecture, hard-working, smiling workers, technology, festivals, vibrant downtown and historic places. There is a reservoir of images used and reused that invoke the themes of progress, culture, nature and business, a world of hard work but with ample opportunity for play, of hard-working people and dynamic industries but friendly atmosphere and relaxed lifestyle, good infrastructure but low taxes, government that helps but does not interfere. The visual images sell a city free from doubt, devoid of conflict.

A vital element in a city's advertising campaign is the slogan. As an eye-catching device, the slogan is one of the simplest and most effective means to implant ideas and to aid in name recognition. City advertising, and advertising in general, has been dominated in recent years by the company slogan, the verbal equivalent of the logo, immediately identifiable, memorable and punchy (Table 9.3). Not all the slogans meet these criteria. Many use the word business (Dallas, Milwaukee, New York, Norfolk, Phoenix), some use the definite article

to describe themselves (Jacksonville, Memphis, Rochester), some stress location (San Jose, Lexington and Memphis) and some are so wacky as to defy categorization. Precisely what positive image Omaha's *Wild creatures loose in the city* is supposed to invoke is difficult to fathom.

The images and slogans both have a repetitive quality but also contain attempts at individuation. Many of the advertisements in the business press involve attempts at distinguishing the city from the rest of the urban system: Boston's rich university tradition; Chicago as a world-class financial center; Rochester, the site of the corporate headquarters of Xerox and Kodak, as the world image center (Figure 9.1); Fairfax's and Oak Ridge's high-tech; Memphis's distribution advantages; Los Angeles' booming economy as the capital of the future (Figure 9.2); Atlantic City's year-round entertainment; Nashville's music; Jacksonville's and Orlando's pro-sports teams; St Augustine's historic heritage; and Miami's seasonal tourism. Every city claims that its particular individual advantage is globally or nationally recognized. The advertisements seek to balance the use of known, familiar and repetitive images and slogans with the employment of distinctive, individual characteristics that reference the distinct advantages and opportunities of specific cities.

Cities marshal a long list of good incentives and images in their ads. To attract and retain business, cities advertise that they possess a pro-business political climate, an ideal workforce, high-technology industries and research institutes, a solid infrastructure and a healthy local economy. Locational advantage is also a selling point for cities located in the Central Time Zone or near bigger cities. Atlanta sells itself as 'strategically located for global business' (Figure 9.3). Many studies on marketing the city note an increased attention to quality of life matters, including healthier, greener environment and cultural, recreational facilities (Burgess, 1982; Goss, 1993). This demonstrates that quality of life has become an important element in the more recent phase of advertising activities. Ward (1994: 58) coins the term 'cultural promotion' to describe cities' appeals to the quality of life. Companies based in local areas is one rhetorical device through which advertisers seek to give credibility to their discourse by referring to their previous successes (Gold, 1994: 30).

Quotations of the opinion of independent surveys on the city are also a popular feature of city advertisements. For example, Kansas City boasts of its number 1 ranking workforce in the US by *Fortune* magazine, based on skills and availability (*Business Week*, 26 December 1994). Fairfax County mentions that last year *City and State* magazine proclaimed the county the best financially managed county in the nation (*Financial World*, 24 May 1994). The testimonial of senior executives of local companies who express their satisfaction with the city is a popular strategy to show how the city works for business.

To fully understand both the reasons and perhaps the relative success of a city's advertising it is appropriate to look at detailed case studies. We will illustrate two cities' marketing stories: Memphis and Milwaukee.

MEMPHIS: AMERICA'S DISTRIBUTION CENTER

Memphis has all the natural assets needed for a distribution center. Its location on the Mississippi River and in the Central Time Zone, and its well-connected highway and railroad systems, indicate that the city meets certain preliminary conditions as a distribution center. Only one additional thing was needed, a nationwide advertising campaign to illustrate Memphis's comparative advantages. In 1981 the Memphis Area Chamber of Commerce, ad agency Walker & Associates and the existing distribution community teamed up to launch the campaign, using magazines and direct mail to spread the word of the city's distribution advantages; they first created the slogan 'Memphis: America's Distribution Center' (Salomon, 1995). The campaign paid for distribution success stories in many magazine ads placed in various business and distribution journals. Local truck companies used the bumper sticker with the slogan on all their trailers. More recently, direct mail marketing to *Fortune*'s 500 companies has taken the form of battery-powered messages announcing 'Good vibrations from Memphis' and a weather radio offering 'A Sunny Forecast from Memphis' (*ibid.*, 1995: 14).

That Federal Express (FedEx) is headquartered in Memphis has played a major role in establishing the city's identity as America's distribution center. A senior executive of the Memphis Area Chamber of Commerce said FedEx gave Memphis's slogan credibility. FedEx was headquartered in Memphis in 1973. At that time, local airport authorities provided incentives to the company, including warehouse and office space (McConville, 1993). FedEx's phenomenal growth in Memphis, which became the community's largest employer, has brought many favorable consequences to the city. David W. Cooley, the president of the Chamber, said, 'We even have a thing called "Memphis-based pricing", where you get a better price from FedEx by being based in Memphis, and you can ship later in the evening' (*Financial World*: 1993, 51).

Three advertisements for Memphis were found in periodicals. Two of them focus on the city's advantages in distribution (see Figure 3.6 in Short and Kim, 1998). The third ad emphasizes the city's first international direct air route: Memphis to Amsterdam. The city called itself 'The New Gateway to the World'. Here we can see how Memphis's advertising, public relations, economic development solicitation and general civic identity all swirl around the city's advantages in distribution.

The Chamber claimed that its marketing programs were responsible for at least 60% of the 103,000 jobs created since 1985 and plans to keep Memphis's national marketing program in place for the next 10 years (Salomon, 1995). It expects a total economic impact in excess of $25 billion on the local economy during that period. As we have seen, Memphis's success story is very clear. It has called itself America's distribution center for over a decade, supported distribution facilities to companies, especially Federal Express, and then successfully attracted other companies.

MILWAUKEE: A SHINING STAR IN THE RUSTBELT

Milwaukee had an image of decline. Its economy has been hit hard by the loss of manufacturing jobs in the past three decades. Milwaukee has suffered from a considerable number of manufacturing plant closures, job losses, and a decaying downtown. *Blue Collar Goodbyes* (Doro, 1992), a collection of poems, is an example of writings that describe devastating consequences of plant closures to communities and workers in Milwaukee.

Milwaukee's current local economic development and city marketing have been greatly driven by its mayor, John Norquist, elected in 1988. A magazine advertisement for Milwaukee (*Forbes*, 19 December 1994; see Figure 3.7 in Short and Kim, 1998) cites Norquist's vision for the city's future: 'Milwaukee should be a private-sector city and a free-market city.' Public–private partnership is his weapon of choice, including the construction of Wisconsin Center in the battle to reinvigorate the city's downtown business district. He gave up completely the city's welfare-state tradition, stating that high taxes and a welfare system had been associated with high crime, family breakdown and neighborhood decay (Eggers, 1993: 70).

In conjunction with many revitalizing efforts, the city decided to launch the magazine advertisement campaign to convey the message of success in ridding itself of its Rustbelt image and attracting business, tourists and residents. 'The City That Works for Your Business' is used as its marketing slogan. Two big advertising supplements in

Advertising Age (20 June 1995) and *Forbes* display the city's continuing economic growth, strong demographics and numerous business assistance programs. These ads also illustrate that Milwaukee's economy is growing and that its unemployment level is lower than the national average. The city compares its figures of population decline with those of other Midwest metropolises in order to assert its better situation. Milwaukee's other attractions to potential residents and tourists are locally based professional sports teams, Bucks (basketball) and Brewers (baseball), lakefront festivals (Summerfest and various ethnic festivals), and Lake Michigan as the destination of family vacations.

Milwaukee's efforts to replace its declining image with a postindustrial image with a vibrant economy and high quality of life, are very similar to the marketing programs of other manufacturing cities like Cleveland and Pittsburgh (Holcomb, 1993), Syracuse (Short *et al.*, 1993) and Glasgow (Boyle and Hughes, 1991, 1995; Paddison, 1993; Loftman and Nevin, 1996), although Glasgow's program has been more culture-driven.

The rise of the new urban entrepreneurialism and the increasing involvement of business coalition in local politics constitute the framework of urban marketing in the US as elsewhere in the world. In magazine advertisements for cities we have noticed the manipulation of place imagery as well as common features that cities try to advertise. The marketing campaign stories of Memphis and Milwaukee provided an introduction to cities' efforts to promote local economic growth by reimagining the city. A number of questions remain. First, how can we assess the effectiveness of urban marketing? There is a real need for a systematic measurement of the effectiveness of these campaigns. Barke and Harrop (1994) cite one officer's skeptical view: 'We do it because everyone else does it.' They comment that many cities are not really aware of any significant direct gain to their localities from their promotional activities. City governments and Chambers of Commerce will always take the most favorable interpretation. The CEO of Coca-Cola once remarked that half of his company's huge advertising budget was a waste of money; the problem was that he did not know which half. Like magic, there is a mystery at the center of advertising. Its precise effect is impossible to measure. Cities have advertising campaigns partly to promote an image to an external audience in order to (re)present or reimagine the city at a time of dramatic changes in the movement of capital and profound shifts in the space-economy at global, national and regional scales. Advertising campaigns are also about shaping and guiding an internal debate. At the International City Conference held in Pittsburgh in May 1993 the three major themes were successful development strategies, utilizing the university as an international linchpin, and what the organizers termed building a local consensus for an international game plan. A standard theme in advertising companies' pitches to cities is the need for programs which improve citizens' belief and pride in their city and promote local allegiance to the municipal marketing strategy. In many cities this involves subordinating welfare goals to the dictates of place competitiveness and the imperative of growth (Logan and Molotch, 1987; Peck and Tickell, 1995).

There is a connection between changes in the space economy and marketing that goes both ways. We have argued that these marketing campaigns have been conducted because of changes in the space economy. But to what extent can marketing campaigns bring about shifts in the space economy? This again poses the difficult issue of measuring the effects and consequences of city marketing. This is an important topic that deserves careful consideration and further work.

While much has been made of city advertising it is important to place it in a broader context. Advertising by cities in the US amounted to only $12 million in the period 1984 to 1987, compared with annual figures of $80 million for Miller Lite and $150 million for Burger King (Bailey, 1989). However, city advertising can afford to be smaller because it is targeted at the business press, the opinion-makers and business leaders. The rank order of advertising dollars spent by publication calculated by Bailey (1989) was *Wall Street Journal, Business Week, Fortune, Forbes, New York Post, Newsweek, Time* and *New York Times*. The effects of this expenditure are impossible to evaluate. Urban advertising campaigns are

also pursued because other cities are doing it. Cities do it because other cities do. In the era of intense place wars, cities need to maintain as high a profile as other cities, if not higher. It is like people in a crowded room trying to catch someone's attention at the front of the room. When just one person stands up the people behind them have to stand just in order to be noticed. In the deregulated 'room' that is the global economic context of the contemporary place wars, all cities need to stand up and be noticed.

Independence of the urban from the national

Globalization processes have given rise to the economic autonomy of cities. It has been argued that city economies are the core of national wealth (Jacobs, 1984), yet gross *national* production (GNP) and *national* growth rate have long been used as main indicators in the estimation of a society's economic power and potential. The rise of the entrepreneurial city, induced by economic globalization and the retreat of national governments in policy making, has been accompanied by cities' efforts to 'delink' or 'decouple' themselves from their respective national economies. Cities with booming local economies, such as Barcelona, project images very detached from their ailing, or at least unimpressive, national economies (Clusa, 1996). Far from being the drags on national competitiveness portrayed in the early 1980s, note Clarke and Gaile (1998), many urban regions have prospered on their own economy in the 1990s.

The heightening importance of cities' independence from the fortunes of their nations' economies is most evident in Europe which is defined as 'a Europe of the regions', presumably urban regions, rather than a Europe of nation-states. Lever (1997) examines a few reasons for recent delinking trends of urban economies in Europe. Cities can outperform their states in terms of economic fortune. Tired of waiting for and begging for their proportionate share of the national output or investment, cities have gradually developed local initiatives for economic development.

The proliferation of city marketing across Europe has encouraged cities to see themselves as distinctive units, instead of being dependent on central governments, and city mayors have acknowledged that they might achieve an international profile by embarking on some economic cooperation with supranational institutions, such as the EU or foreign cities.

Cross-border cooperation and trans-frontier networking involving city authorities in Europe is a growing phenomenon. Through a case study of recent developments in cross-border cooperation involving cities in Britain and France, Church and Reid (1996) examine the implications of such transnational cooperation on urban politics. Unlike the traditional state-centric foreign relations, this new form of cooperative program allows city officials to negotiate with representatives of the European Commission and with other foreign cities. Although urban governments still have less power than their national counterparts, Church and Reid (1996) argue that bottom-up local economic strategies, such as EU-supported cross-border projects, have certainly contributed to the restructuring of the state and central–local relations. The growing involvement of urban governments in transnational cooperation and networking has given rise to both the supranational institution (the Europeanization of national policy making) and urban authorities.

The global–urban nexus

The literature on the relationship between globalization and urban changes has produced a wealth of case studies that exemplify the importance of local context and initiatives in economic development. Avoiding a unidirectional determinism from global to local, case studies point out local factors, agencies and initiatives in mediating the global–urban relationship.

The relationship between globalization and urban dynamics has been taken up by a great number of case studies on contemporary urban changes. Along with world cities researchers, numerous commentators in the fields of locality studies and structural (re)adjustment have addressed the importance of linking different geographical scales ranging from the global to the local in the analysis of restructuring at the urban and regional level (Beauregard, 1995; Lipietz, 1993; Swyngedouw, 1997). As we have seen in previous sections, social, economic and political relations and processes in the city are increasingly tied to forces operating at a global scale. The global–local connection has become a highly valued framework to examine the new economic, cultural and political realities that contemporary cities are experiencing.

Case studies reveal that not all cities are affected equally by globalization. Each city has its particular interplay of the global and the local, as specific local traditions in economic, cultural and political terms actively rework global forces. Beauregard (1995) identifies five major points in the global–local connection. First, the dominant global forces are economic. Second, these forces penetrate to the local scale in an uneven fashion and even bypass certain institutions, industries, people and places. Third, global forces are sometimes embraced, sometimes resisted, and sometimes themselves exploited. Fourth, economic, political and social forces operate at a variety of spatial scales: global, national, regional, urban and neighborhood, for example. Finally, and as a result, because each scale is relatively autonomous from the others, global forces are mediated as they penetrate downward.

Because of the globalization of production, consumption and exchange in association with the decline of national regulation in wage fixing and work practices, the 'national' becomes a less significant unit of analysis. The 'global' becomes more significant as the unit for understanding general economic trends (see Reich, 1991) while the 'urban' becomes the unit for appreciating the intersection between capital and labor, economy and society, polity and comity.

Resistance, embrace, submission and accommodation of individual cities to the global has been acknowledged and asserted, if not carefully examined. For a better understanding and theorization of the global–local connection we need more detailed biographies of cities within a global context. Few, however, propose any specified framework to examine and measure urban changes influenced by globalization processes. Major research foci identified by world cities research, such as the concentration of multinational firms' headquarters, the fast rise of financial and producer services and the presence of advanced telecommunications infrastructures, may not be adequate for explaining urban dynamics elsewhere, particularly in non-Anglo-American cities in the developing world. Most researches on the relationship between globalization and urban change have also shown a strong orientation to the urban impact of 'economic' globalization, missing out the cultural and political dimensions of globalization and their urban outcomes. By looking at the general rise of the entrepreneurial city and also particular entrepreneurial cities we may be in a stronger position to connect, both theoretically and empirically, the global–urban nexus, the economic with the cultural and the political.

10

Sydney: Going for Gold in the New Australia

The 2000 Summer Olympic Games may have seemed a long way off in 1996. Not in Sydney, Australia, where a giant clock in the airport entrance hall and others around the city counted down the number of weeks long before the 2000 Games began.

The city was gearing up to be the host site for the Summer Olympics in 2000 as early as 1994. The Olympic logo, a combination of the outlines of a boomerang, the Sydney Opera House and a figure running, was printed on millions of T-shirts, hats, mugs, posters and tickets. New stadia and transport links were constructed on a former toxic waste site in Homebush and the whole public talk of the city in the last four years of the millennium was dominated by the Olympics. Car registration plates had the logo 'Towards 2000'. The *Sydney Morning Herald* had a weekly feature entitled 'Olympic City' that solicited ideas to enliven Sydney in the run-up to the Games as well as daily coverage of the funding, organization and management of the Games. Local television news broadcasts carried regular reports from SOCOG (Sydney Organising Committee for the Olympic Games) as well as breaking stories of emergencies, crises, incompetences, failures and successes that go with organizing any global event. The city was in the grip of a rising Olympics fever as the clocks ticked down to zero.

The Olympic Games have become important, not just as sporting events but as massive urban renewal programs and image make-overs. The Games are a global spectacle that put the host city at the center of world attention. For urban boosters around the world becoming the host of the Games is the ultimate prize in name recognition

and international respectability. The Games allow cities to change their global ranking and identity; they allow the possibility of the image of the city to be transformed from anonymous, provincial hicktown to happening global place.

In Sydney the millennium Games mark the culmination of a gradual shift in identity from imperial outpost of the British Empire to Australia's global city.

Rewriting the city, refashioning national identity

Australia was founded by the British as a convict colony, an antipodean gulag. The American Declaration of Independence meant the British lost a convenient dumping ground for convicted criminals. The voyages of Captain Cook had uncovered the existence of a southern landmass. And so it was that in January 1788 Captain Arthur Phillip's little fleet of two warships, one sloop, three supply ships and six ships carrying 778 convicts sailed up to what was named as Sydney Cove.

British control eventually extended to the whole continent. Free settlers replaced convicts and the country became less of a gulag and more an economic branch plant of 'British Empire, Inc.' headquartered in London. Throughout the nineteenth century, a series of self-governing states – New South Wales, Queensland, South Australia, Western Australia and Victoria – looked more to London than to each other. The railroad system linked the interiors to the respective capital cities rather than to each other. Raw materials were collected from the interior, shipped to the coastal ports and sent to Britain. In return, Australia received manufactured goods and an identity.

It was only in 1901 that the individual states federated to become Australia. State rivalries, especially between the two most populous states, New South Wales and Victoria, were so intense that a brand new capital was built on a neutral site. Canberra, in contrast to all the other major Australian cities, sits alone on the interior, a reminder of the inability of Melbournites to accept Sydney, and of Sydneysiders to approve of Melbourne, as the capital of the new country.

Throughout the first half of the twentieth century, Australia was politically independent but culturally dependent. The head of state was, and still is, the British monarch; up until the later 1940s Australians were British citizens; and immigration policy kept out everyone except white Anglo-Celts. Australia was Britain under a southern sky.

The change was gradual. After the Second World War a severe labor shortage meant that Australia encouraged immigration from Southern and Eastern Europe. Slavs, Italians and Greeks made the long sea crossing as well as those from Glasgow, Birmingham and Newcastle. Foreign languages began to be heard and the stodgy British cuisine was enlivened by Mediterranean flavors.

In 1973 Britain joined the Common Market. Part of the entry price was a jettisoning of Britain's special economic relationship with Australia as well as New Zealand. Wool, beef and lamb from the antipodes could no longer be given preferential treatment in the British economy. The former colonies had to fend for themselves in world markets. The cutting of old economic ties fostered the establishment of new cultural identities.

The pace of change picked up in the 1970s. A more culturally independent Australia was emerging. Ambitious Australians no longer felt the same compulsion to move overseas to make it on the world stage, and the cultural cringe, a deference to all things British, was replaced by the cultural strut. The election of Gough Whitlam's Labour Party, with its more radical agenda, in 1972 marked a political watershed. Stirrings of Republicanism were now getting stronger, many no longer wanting the Queen of England as their head of state, her portrait on all coins and the Union Jack in the top left-hand corner of the Australian flag. Novelists such as Patrick White, Thomas Keneally and David Malouf, painters such as Fred Williams and Brett Whiteley, and film-makers like Bruce Beresford and Peter Weir gave voice and image to a more confident and culturally independent Australia. Sydney was at the center of this cultural renaissance.

Many Australians began to define their country less as an antipodean version of Britain and more as occupying a space between the old superpower of Britain and the new superpower of the USA. Compared to Britain, Australia had the same civility without the stuffiness and the crippling class consciousness. Australia was a 'newer' country like the USA, with the same sense of limitless possibilities, but softer, less violent.

By the beginning of the 1980s a new Australia was being fashioned – no longer neo-British but more multicultural, and no longer a UK branch plant but more a member of a Pacific Rim economy. National identity draws upon the shared experience of citizens. In the period 1962–66, 50% of immigrants came from the UK, 11% from Greece and 10% from Italy. In the period 1992–96, of the one million immigrants only 12% came from the UK and 22% came from Asian countries including China, Hong Kong, Vietnam and the Philippines. These 'new Australians' added a new element to the overwhelmingly white European 'older Australians'.

National identity is never fixed. There is an economic bedrock to changing identity. In the seismic shift following Britain's entry into the Common Market, Australia looked to markets in Asia-Pacific. Japan is now the biggest single importer of Australian exports and over 60% of all Australia's exports go to Asia. In the boom years the Asian connection seemed to promise unending growth and expansion; in the last 18 months it has meant recession and contraction.

Sydney is at the heart of these changes. The city is a major destination point for Japanese, Korean and Chinese tourists. The white working-class district of Cabramatta has become an Australian-Vietnamese suburb. The whole cul-

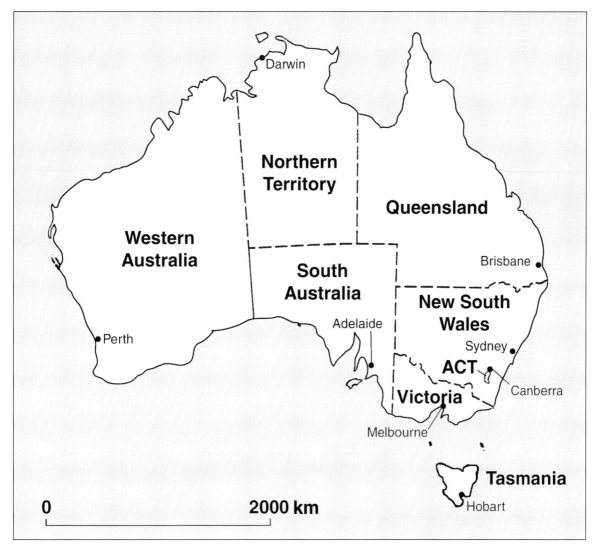

Figure 10.1 Map of Australia.

tural economy of the city is more Pacific Rim. Trucks still drive on the left-hand side of the road on their way to the docks but the ships are increasingly carrying more goods to and from Asian-Pacific countries.

Becoming Australia's global city

Sydney beat a number of cities to become the host of the 2000 Olympics, including Beijing, another city in a country also seeking to change its international image. But Sydney's main competition was not from overseas, it was closer to home; and it was not just for the Games, it was to reaffirm its recent status as Australia's global city.

For almost 200 years the economic if not the cultural history was less of a single nation and more of a series of city states. Each of the state capitals – Adelaide, Brisbane, Melbourne, Perth and Sydney – functions as the center of a powerful state (Figure 10.1). In a global economy, however, small countries can support only one

Figure 10.2 Sydney Opera House: the setting. (*Photograph by John Rennie Short*)

Figure 10.3 Sydney Opera House: detail. (*Photograph by John Rennie Short*)

global city, the place of major corporations, financial services and cultural industries. The main competition throughout this century was between Melbourne and Sydney. Melbourne had a lot of advantages; it was the temporary capital until the government moved to Canberra and it maintained a reputation for a vibrant intellectual life as well as a flourishing economy. Melbourne hosted the 1954 Olympic Games. Since then, however, Sydney has eclipsed its former rival as *the* Australian city. Ask people to think of an image of Australia and most people around the world will mention, along with the kangaroos and Ayers Rock, the Sydney Opera House and maybe Sydney's Harbour Bridge. Melbourne does not have an icon of comparable international recognition.

The Sydney Opera House sits on Bennelong Point, jutting out into the harbour (Figure 10.2). It used to be a tram shed. The New South Wales State Government announced an international competition for the design of an opera house in 1955. The winning entry was from the Danish architect Joern Utzon. The tram shed was demolished and construction began in 1959. It took eight years and final costs were 15 times the original estimate, before the building was completed; but the final result was worth the wait and cost. The shell-like structure (Figure 10.3), also looking like sails from a distance, gave Sydney not only an opera house, but one of the great buildings of the twentieth century and an immediately recognizable image for the city. The outline of the Opera House appears in the logo of the 2000 Summer Olympics.

Sydney has been lucky with urban projects. After the Opera House came the redevelopment of Darling Harbour. This old shipping terminus became a successful inner city redevelopment project with aquariums, museums, shops, restaurants and a casino. Sydney seemed to have the golden touch, helped, no doubt, by its wonderful physical setting, its beautiful natural harbor and a salty clear light that casts everything in a flattering glow.

The urban archetypes that circulate within the country have also reinforced Sydney's premier position. In Australia, Sydney is a city of sunshine, beaches, a subtropical hedonism of bodily pleasures. Melbourne is a city of interiors, grey skies, grid-pattern streets and trams. In the dualisms that constitute the national consciousness, Sydney has become a place of pleasure, the city of the body; Melbourne a place of the mind, the cerebral city. Sydney has become the Australian city, the urban point of contact between the country and the rest of the world. Brisbane and Melbourne bid to host the Games in 1992 and 1996 respectively. Only Sydney's bid was successful. The Olympics will reinforce Sydney's prime position in the Australian urban hierarchy and confirm its global recognition.

Compared to the Olympic Games in Los Angeles and Atlanta, the state rather than private corporations will play a lead role. The Sydney Games have a rather old-fashioned big-government emphasis. This is understandable in the Australian context where the state governments have wielded enormous power since before Federation. In Captain Phillip's first fleet, there were almost 300 marines and civil servants. Gaolers, government officials and a severe regulation system landed with the British as well as recalcitrant criminals. Extensive state power has been wielded ever since and, like all institutions, state governments have had difficulty giving back power to the federal government, the people and the market. These Games have been promoted, planned and paid for by the New South Wales Government. They are meant as a showcase for the capital of the state. Big government may be out of fashion but it is alive and well in New South Wales and, in keeping with an increasingly competitive world, profoundly entrepreneurial. This public emphasis suited the Olympic Committee who awarded the Games to Sydney because they were disturbed at the creeping commercialism of the Games and especially the 'corporate games' in Atlanta.

Reconciliation: a new history

As Sydney and Australia work diligently to create a new identity for the future, they are also grappling with the past. The creation of a new identity for both Sydney and Australia has involved looking back as well as forward. The traditional history of the city and country passed over the indigenous inhabitants. The oldest Australians were ignored in the creation of the neo-British identity; they were seen as remnants of a stone-age culture quickly passing into oblivion. They had no official place in the new democracy. Even to this day, foreign visitors to Sydney see little significant Aboriginal presence in the multicultural life of the city.

At the time of white contact in 1788 there were approximately three quarters of a million Aborigines. Their subsequent history is a sad and dispiriting tale of loss of land, loss of rights, loss of a place in the emerging nation. Their land was taken by the crown and then sold or leased to settlers. Aborigines were herded into a new version of a gulag and denied such basic rights as the right to vote or even to be counted in the census. White Australia turned its back on the original inhabitants.

In recent years reconciliation is replacing condescension. The historian Henry Reynolds has documented the terrible treatment of Aborigines. In *The Other Side of the Frontier* (1981), *Fate of a*

Free People (1995) and *Aboriginal Sovereignty* (1996) he has recounted a bleak tale. A more recent book, *This Whispering in Our Heart* (1998), profiles nineteenth-century white Australians who championed the Aboriginal cause. It is a book that echoes contemporary concerns.

LAND RIGHTS AND NATIVE TITLE

Land has been central to the story of Aborigines since 1788. The basic narrative, as in North America, was that the indigenous people had it and the Europeans took it from them. Land was not only a source of sustenance for the indigenous people, it was of tremendous cultural and spiritual significance. Losing the land meant losing not only your economic base, but also your culture, your cosmology, your place in the universe.

The successful fightback came late in Australia. It was only in 1967 that Aborigines were given citizenship in Australia. The first significant measure, undertaken by the Whitlam Government, was the 1976 Aboriginal Land Rights Act which recognized the land claims of indigenous communities. Under this measure the land around Ayers Rock (Uluru) and the Olgas (Kata Tjuta) was successfully claimed by the indigenous Anangu community; in 1985, it was subsequently leased back to the government as the Uluru–Kata Tjuta National Park. Land rights legislation has now been passed in all states except Western Australia. The legislation introduced into the Australian legal system complex anthropological issues of how to define traditional land rights. Primary spiritual responsibility and cultural attachment had to be proven for land rights claims to be successful. Anthropologists became legal witnesses and painting, songs and myths became evidence of property claims.

Land rights affected only public (Crown) land that was not leased. Claims elsewhere were impossible under the existing legal arrangement which was based on the assumption of *terra nullius*. The British Crown claimed all land in Australia without recognition of prior ownership. No treaties were signed, no payments were made. *Terra nullius* was reconfirmed from 1788 to 1982 with ownership transferred from the British Crown to the federal government and the individ-

ual states in 1901. All this changed in 1982 when three Torres Strait Islanders, Eddie Mabo, David Passi and James Rice challenged the concept. On 3 June 1992 the High Court ruled that *terra nullius* should not have been applied to Australia and native title to land did exist. Two of the judges, Justices Deane and Gaudron, wrote of Australia's treatment of Aborigines as a 'national legacy of unutterable shame'. It was not an out-and-out victory as the judgment was vague in places and allowed the extinguishment of native title. A Native Title Act in 1993 sought to clear up some of the thorny issues. This legislation also established a land fund, under the Indigenous Land Corporation (ILC), for the purchase of land by indigenous communities. A\$4 billion was allocated for land acquisition to the year 2004. By June 1998 the ILC had obtained 39 'culturally significant' properties at a cost of A\$36 million, with approvals for another 96 properties at an expected cost of A\$94 million. One of the purchased properties was the 63,000-acre Mogila Station on the NSW/Queensland border. Famous for its merino rams, the Station will be owned by the ILC through a subsidiary and gradually transferred to the indigenous community.

Much of the interior of Australia is leased rather than owned. Away from the fertile coastal plain, the vast bulk of inland Australia is Crown land that has been leased to pastoralists and miners. The leasehold system was developed in the 1830s to avoid pastoralists from illegally working and claiming Crown land. The system allowed access to land for the farmers and gave the government a source of revenue as well as ultimate control. The system is now regulated by the states. The Mabo ruling and the 1993 Act did not say anything about leasehold land. In 1993 the Wik peoples began legal proceedings to claim native title to land in Cape York Peninsula, Queensland. There were currently 10 pastoral leases in the area, but the Wik peoples stated that the granting of leases, in line with the Mabo ruling, did not extinguish their title. The case made its way to the High Court, which in December 1996 ruled that granting of leases did not grant exclusive possession and did not extinguish native title. The decision allowed the coexistence of pastoral lease and native title.

The Wik decision was seen by pastoralists and mining interests as an attack on their rights. The National Farmers Federation and mining interests inflamed white fears of an Aboriginal take-over. The *Bulletin* magazine began a story in July 1997 with the line, 'Development of the largest agricultural project in north-west Australia is paralyzed by a native title claim.' The sentence captures the fear that growth will be halted by greedy Aborigines. The Wik decision was a political minefield.

In 1996 the (more right-wing) Liberal Party in coalition with the rural-based National Party won a landslide election. The more urban-based Labour Party, which had little support from white rural Australia, and could more easily promote native title, lost its long hold on power. The new Prime Minister, John Howard, refused calls from his backbenches and the states for the extinguishment of native title on pastoral leases. In 1997 a government plan emerged. It was a compromise that stated that native title can be extinguished to the extent that it is inconsistent with the rights of pastoralists, but can coexist to the extent that the rights are not inconsistent. Meant to offend no-one, it displeased everyone. The plan was formalized as a bill and meandered a tortuous way through the federal government and national public debate. The land issue which had begun as a source of reconciliation ended up as a major point of conflict.

THE STOLEN GENERATION

In the early 1990s reconciliation was in the air. The Labour Government established a Council to work towards Aboriginal Reconciliation by the year 2001, 100 years after Federation. In 1995, the Attorney General called for an inquiry into the treatment of Aboriginal children, who for over 100 years had effectively been wards of the individual states. The inquiry published its report, *National Inquiry into Separation of Aboriginal and Torres Strait Islanders' Children from Their Families*, in Sydney in May 1997. It caused a sensation. Here were heart-rending tales of children being forcibly removed from families to be brought up in government institutional homes or raised by white families. For most of this century

right down to the 1970s, many Aboriginal children were forcibly removed from their families by the state. The report estimated that 10% of Aborigines over 25 had been removed from their families. In some districts it was closer to 30% and even 50%. While there were many individual acts of white kindness, the system was a form of cultural genocide for Aboriginal identity. The effects on the children were devastating. Those separated were three times as likely to go to prison and twice as likely to take banned substances than those not separated. This was not a dry government report but a careful examination and vivid description of human suffering. Land issues could get lost in legal technicalities and obscure Latin phrases, but the separation of children from their families was a compelling and moving story that stirred the conscience of the nation. A National Sorry Day was declared, not by government but by grass-roots mobilization. On 26 May 1998 over 300 events were held throughout the country. In Sydney, in a symbolic act of reconciliation, a member of the stolen generation, Mrs Nancy de Vries, received an official apology from the Governor of New South Wales, Sir Gordon Samuels, and thousands of schoolchildren in the city donned the Aboriginal colors of red, black and yellow in a demonstration of support and atonement. In *Sorry Books* provided in the city and around the country, over a million people signed their names, often appending heartfelt messages of apology and reconciliation. It was one of Australia's finest moments and perhaps a model for other countries seeking to deal with histories and present realities of genocide and dispossession.

In the spirit of the new history, the recently constructed Museum of Sydney gives due recognition to the original inhabitants of the area and provides a *Sorry Book* for those willing to sign and leave a message (Figure 10.4).

The backlash

Changes in national identity are never easy or sudden. Societies, groups and individuals invest emotional capital into their 'national' sense of

Figure 10.4 Sculptures outside the Museum of Sydney. (*Photograph by John Rennie Short*)

themselves. Too sudden and rapid a devaluation of cultural currency creates tensions and uncertainties. Into the intersection of the redefinition of national identity, Aboriginal issues and a stagnant economy has emerged the One Nation Party. It was launched in Queensland at the Ipswich Town Hall on 11 April 1997 by Pauline Hanson. She had made a reputation as a recently elected federal Member of Parliament who, in her inaugural speech the year before, had described an Australia 'swamped by Asians' and spoke of a 'taxpayer-funded Aboriginal state'. The One Nation Party was against Native Title, immigration, welfare payments and restrictions on guns. Over the next year, she and her party were vilified in the press. She was seen as an embarrassment to a country dependent on Asian investment and tourism and a city about to host the Olympic Games. She was routinely described as demented, part of the tiny lunatic right. All this changed on Saturday 13 June 1998. On that day state elections were held in

Queensland. One Nation captured 23% of the votes and gained 11 of the 89 seats in the state parliament. Some of this could be written off as a Queensland effect, which in Australia occupies the same mythic space as Mississippi or Alabama does in US racial ideology. The state has had a series of right-wing governments and a history of abysmal treatment of Aborigines. Its racist attitudes allowed it to be regularly described as the Deep North. But there was evidence of wider support. Opinion polls taken after the election gauged a 20% support in rural New South Wales and even higher in Western Australia.

Pauline Hanson is an unlikely figure to break the mold of Australian politics. She is a twice-married mother of four who has a series of well-documented relationships with men. Her previous experience was as owner of a seafood take-out – the image of a fish-and-chip lady was burnished by one of her early handlers – who had amassed enough money for a comfortable living.

Figure 10.5 Braidwood: a country town in New South Wales. (*Photograph by John Rennie Short*)

She was not a professional politician, and it showed. She was neither smooth nor knowledgeable. When asked questions on national television she would sometimes say candidly. 'I don't know'. Pauline Hanson gained political support by not being a traditional politician. Like Ross Perot her amateur status was part of her appeal.

Discontent with politicians is endemic in all societies. It is doubtful if Australia has more corrupt politicians than anywhere else, but sometimes this seems to be the case as a constant succession of state and federal politicos are discovered to be claiming phoney expenses. In 1997 three federal ministers were forced to resign, caught cheating their expense forms, and in 1998 a federal MP, Noel Crichton-Smith, was fined A$8,000 for the same offence. On the scale of human venality it is all low-grade stuff, but in the middle of a national redefinition it further corrodes the bond between governed and governing.

The One Nation Party was gaining support in rural areas at the expense of the National Party, the junior partner with the Liberals in the Federal Coalition Government. In the bush, as the rural areas are called in Australia, conditions have been worsening as economic rationalization works its way across the vast landscape. Banks and post offices have closed in many small towns; unemployment rose from 7% to 12% in 1997–98 while it fell in Sydney to 7% (Figure 10.5). National unemployment rates have been hovering around 8% for the past few years. Income levels in the bush are a third of what they are in the city. The same globalization that made Sydney a hub in the international financial chain of command also made economic black holes of small country towns. The people of the bush felt under threat from Native Title, deregulation of public utilities and the widespread abandonment of public services and private capital in the wake of economic rationalism.

Economic rationalization, a sort of antipodean mix of Thatcherism and Reagonomics, was a policy of deregulation and letting the market decide, that has been in operation in Australia since early 1980. There had been a bipartisan consensus that the economy was central to federal government deliberations and that civil society had to get in line. The sociologist Michael Pusey, in his 1992 book *Economic Rationalism in Canberra*, wrote of an antisocial internationalism that was the dominant mindset in Canberra shared by senior bureaucrats and powerful politicians. Economic competition, tariff reduction, deregulation and letting market forces decide social outcomes, were the themes and policies of this ideology. Economic rationalism had some of its most antisocial effects in the bush.

Support for One Nation was a spasm of resentment against the redirection of national identity, the worsening economy and the consequences of economic rationalism. One Nation offered a return to a simpler time, the days of a White Australia, an economically secure Australia, a white Dreamtime of ethnic solidarity, cultural security, steady growth and full employment. A television interviewer, clearly exasperated at Hanson's simplistic attitudes, told her 'You can't take us back to the 1950s.' 'Why not,' she coolly answered, 'there was nothing wrong with the 1950s.'

One Nation also offered an easy refuge for the resentment. I visited Cairns, far into the Deep North of Queensland, one week after the election. I spoke with a businessman about why people voted for One Nation. 'Asian immigration and Aborigines', he replied immediately. He was not racist, he assured me, but believed that there was too much Asian immigration. He had felt unable to speak beforehand because of political correctness. This may explain the underestimation of support for Hanson's party in opinion polls before the anonymity of the Queensland election. 'Aborigines', he went on to say, 'received too much public support.' He informed me that each Aborigine received A$200,000 each year in government handouts. What also came out from many of my interviews was a stifling sense of being unable to discuss these topics. Many of those who voted for One Nation were not political Neanderthals or neo-fascists. From my discussions I came away with an overwhelming sense of people unable to discuss fundamental issues about the direction of their country. One Nation picked up some wacko support but it also claimed the votes of those disaffected with the political process, upset with the economic rationalist agenda and challenged by the changes in Australian identity.

Australia is a tolerant society, but one where conformity prevails. Political discussions are rancorous, but rarely deeply divisive. There is not the same cacophony of voices on any issue that one finds in the US. There is not the same devotion to free speech. Unlike the US where it seems as if no issue is ever agreed upon, a national consensus operates in Australia. The White Australia policy, economic rationalism and new Pacific Rim identity were imposed on civil society rather than emerged from it. There are few national conversations in Australia. The Queensland election prompted the beginnings of one. An Australian of Asian descent wrote to the *Sydney Morning Herald* immediately after the Queensland election: 'In recent years, political correctness has hidden or suppressed not only the most overt forms of racist, but has also labelled the mere questioning or discussion of the other groups as racist ... now we have a valuable opportunity to discuss who we are as Australians, which is included and excluded and what kind of community we want to live in'. Not everyone was so sanguine. I spoke with a Korean-Australian who had lived in Australia for 11 years. She perceived high unemployment and Pauline Hanson's One Nation as a dangerous mixture that posed a real threat to her and her family.

The Queensland election result caused a reassessment and the beginnings of serious national debate. What had gone wrong in the bush? Why had governments not listened? How could a multicultural Australia be defended and celebrated rather than imposed and assumed? There was also an immediate political fall-out. The Coalition Government addressed issues likely to appeal to the bush, including the promise of no post office closures and support for a A$8 billion railway from Melbourne and Adelaide to Darwin. The government also passed a compromise Wik bill in early July 1998, thereby avoiding a dissolution of federal Parliament and the possibility of handing One Nation the opportunity for more electoral success. The independent senator who had previously opposed the government, Brian Harradine of Tasmania, held his nose and voted with the government to avoid what he saw as the real possibility of a socially divisive, race-based election. The bill is vague enough to allow litigation for years ahead.

Cleaning up for the Games

In 1954, 3,000 participants from 67 countries visited Melbourne and a White Australia that prided itself on its ties with Britain and its economic security. In the year 2000, the spotlight of global television coverage will turn its attention to a world city and a multicultural society that will be the host to over 10,000 athletes from over 200 countries as well as hundreds of thousands of visitors. Both the city and the country are as much in the process of becoming as being; they are becoming more multicultural, more Pacific Rim, more globally connected. As we have seen, this redefinition is not without its difficulties, tensions and ambiguities. National reconstruction never is: issues of ethnic identity, the role and place of immigration, a growing inequality as globalization creates winners and losers within the same national borders, an undercurrent of racist feeling often hidden by political correctness, are concerns as central to the new Australia and the emerging Sydney as they are to the wider world.

The main venue of the Games will be Homebush, an inland site that had become a toxic dumpsite. Most of the Olympic Games will take place on this brownfield site. The toxic legacy is being cleaned up, ready both for use in the Games and for a life beyond the Games as the largest sports complex in the state and as the newest residential district in the city. In a material sense, the coming of the Games has allowed a clean-up of the past and something new for the future. The site of the 2000 Summer Olympic Games also provides a compelling metaphor for both Sydney and Australia as well as the wider world: the promise of a revisited past and a new future, as we approach the end of the second millennium.

References

Abbott, Carl, 1996, 'The internationalization of Washington, D.C.', *Urban Affairs Review*, **31** (5), pp. 571–594.

Abler, Ronald F., 1991, 'Hardware, software, and brainware: mapping and understanding telecommunications technologies', in Stanley D. Brunn and Thomas R. Leinbach, eds, *Collapsing Space and Time: Geographic Aspects of Communications and Information*, London: HarperCollins Academic, pp. 31–48.

Abu-Lughod, Janet L., 1995, 'Comparing Chicago, New York, and Los Angeles: testing some world cities hypotheses', in Paul L. Knox and Peter J. Taylor, eds, *World Cities in a World System*, Cambridge: Cambridge University Press, pp. 171–191.

Advertising Age, 1997, '100 years of advertising in New York', 27 January, pp. C1–C18.

Advertising Age, 1998, 'Top 10 agencies by U.S. media billings', 27 April (http://adage.com/dataplace/archives/dp224.html).

Advertising Age, 1998, 'World's Top 50 global ad organizations', 27 April (http://adage.com/dataplace/archives/dp226.html).

Alger, Chadwick, 1990, 'The world relations of cities: closing the gap between social science paradigms and everyday human experience', *International Studies Quarterly*, **34**, pp. 493–518.

Alles, Patrick, Adrian Esparza and Susan Lucas, 1994, 'Telecommunications and the large city–small city divide: evidence from Indiana cities', *Professional Geographer*, **46** (3), pp. 307–316.

American Banker, 1995, 'Foreign banks in the United States', 28 February.

Amin, Ash and Nigel Thrift, 1992, 'Neo-Marshallian nodes in global networks', *International Journal of Urban and Regional Research*, **16** (2), pp. 571–581.

Amin, Ash and Nigel Thrift, 1994, 'Living in the global', in Ash Amin and Nigel Thrift, eds, *Globalization, Institutions, and Regional Development in Europe*, Oxford: Oxford University Press, pp. 1–22.

Andersen, Kurt, 1997, 'A city on a hill: the new Getty Center is a triumph of nineteenth-century ambition,' *The New Yorker*, 29 Sep, pp. 66–73.

Appadurai, Arjun, 1990, 'Disjuncture and difference in the global cultural economy', *Theory, Culture and Society*, **7** (2/3), pp. 295–310.

Appadurai, Arjun, 1996, *Modernity at Large: Cultural Dimensions of Globalization*, Minneapolis: University of Minnesota Press.

Ashworth, G. J., 1998, 'The conserved European city as cultural symbol: the meaning of the text', in Brian Graham, ed., *Modern Europe: Place, Culture and Identity*, London: Arnold, pp. 261–286.

Ashworth, Gregory J. and Henk Voogd, 1990, *Selling the City: Marketing Approaches in Public Sector Urban Planning*, London: Belhaven Press.

AT&T, 1995, 'Korea Telecom introduces WorldSource Services' (http://www.att.com/press/1095/951030.bsb.html).

Auer, James, 1995, 'The mayor who preaches design', *Progressive Architecture*, May, pp. 92–95.

Bagchi-Sen, S., 1997, 'The current state of knowledge in international business in producer services', *Environment and Planning A*, **29**, pp. 1153–1174.

Bailey, J., 1989, *Marketing Cities in the 1980s and Beyond*, American Economic Development Council: Cleveland State University Press.

Barber, Benjamin R., 1995, *Jihad vs. McWorld*, New York: Ballantine Books.

Barke, Michael and Ken Harrop, 1994, 'Selling the industrial town: identity, image and illusion', in John R. Gold and Stephen V. Ward, eds, *Place Promotion: The Use of Publicity and Marketing to Sell Towns and Regions*, Chichester: John Wiley and Sons, pp. 93–114.

Barnet, Richard and John Cavanagh, 1996,

'Homogenization of global culture', in Jerry Mander and Edward Goldsmith, eds, *The Case against the Global Economy and for a Turn toward the Local*, San Francisco: Sierra Club Books, pp. 71–77.

Bassett, K., 1993, 'Urban cultural strategies and urban regeneration', *Environment and Planning A*, **25**, pp. 1773–1788.

Bassett, Keith, 1996, 'Partnerships, business elites and urban politics: new forms of governance in an English city?', *Urban Studies*, **33** (3), pp. 539–556.

Batten, David F., 1995, 'Network cities: creative urban agglomerations for the 21st century', *Urban Studies*, **32** (2), pp. 313–327.

Batty, M., 1991, 'Urban information networks: the evolution and planning of computer-communication infrastructure', in John Brotchie, Michael Batty, Peter Hall and Peter Newton, eds, *Cities of the 21st Century: New Technologies and Spatial Systems*, New York: Longman Cheshire, pp. 139–158.

Baum, Scott, 1997, 'Sydney, Australia: a global city? testing the social polarisation thesis', *Urban Studies*, **34** (11), pp. 1881–1901.

Beauregard, Robert, 1991, 'Capital restructuring and the new built environment of global cities: New York and Los Angeles', *International Journal of Urban and Regional Research*, **15** (1), pp. 90–105.

Beauregard, Robert A., 1993, *Voices of Decline: The Postwar Fate of US Cities*, London: Blackwell.

Beauregard, Robert, 1994, 'Capital switching and the built environment: United States', *Environment and Planning A*, **26**, pp. 715–732.

Beauregard, Robert, 1995, 'Theorizing the global-local connection', in Paul L. Knox and Peter J. Taylor, eds, *World Cities in a World System*, Cambridge: Cambridge University Press, pp. 232–248.

Beaverstock, Jonathan V. and Joanne Smith, 1996, 'Lending jobs to global cities: skilled international labour migration, investment banking and the City of London', *Urban Studies*, **33** (8), pp. 1377–1394.

Beaverstock, J. V., H. Lorimer, R. G. Smith, P. J. Taylor and D. R. F. Walker, 1998, 'Relational studies of world cities: measurement methodologies', GAWC Research Bulletin, **2**.

Behrman, Jack and Dennis Rondinelli, 1992, 'The cultural imperatives of globalization: urban economic growth in the 21st century', *Economic Development Quarterly*, **6** (2), pp. 115–126.

Bergesen, Albert, 1991, 'Semiotics of New York's artistic hegemony', in Resat Kasaba, ed., *Cities in the World-System*, New York: Greenwood Press, pp. 121–134.

Berner, Erhard and Rüdiger Korff, 1995, 'Globalization and local resistance: the creation of localities in Manila and Bangkok', *International Journal of Urban and Regional Research*, **19** (2), pp. 208–221.

Boniche, Armando, 1998, 'Deconstructing Miami's global habit', unpublished paper, Department of Geography, Syracuse University.

Booth, Peter and Robin Boyle, 1993, 'See Glasgow, see culture', in Franco Bianchini and Michael Parkinson, eds, *Cultural Policy and Urban Regeneration: the West European Experience*, Manchester University Press, pp. 21–47.

Bosman, Jeroen and Marc de Smidt, 1993, 'The geographical formation of international management centres in Europe', *Urban Studies*, **30** (6), pp. 967–980.

Bowen, Jr. T. and T. Leinbach, 1995, 'The state and liberalisation: the airline industry in the East Asian NICs', *Annals of the Association of American Geographers*, **85**, pp. 168–193.

Boyer, M. Christine, 1992, 'Cities for sale: merchandising history at South Street Seaport', in Michael Sorkin, ed., *Variations on a Theme Park: The New American City and the End of Public Space*, New York: Hill and Wang, pp. 181–204.

Boyer, Robert and Daniel Drache, eds, 1996, *States against Markets: the Limits of Globalization*, London: Routledge.

Boyle, M., 1997, 'Civic boosterism in the politics of local economic development – "institutional positions" and "strategic orientations" in the consumption of hallmark events', *Environment and Planning A*, **29** (11), pp. 1975–1997.

Boyle, Mark and George Hughes, 1991, 'The politics of the representation of the "real": discourses from the Left on Glasgow's role as City of Culture, 1990', *Area*, **23**, pp. 217–228.

Boyle, M. and G. Hughes, 1995, 'The politics of urban entrepreneurialism in Glasgow', *Geoforum*, **25**, pp. 452–469.

Braman, Sandra and Annabelle Sreberny-Mohammadi, 1996, *Globalization, Communication and Transnational Civil Society*, Cresskill, NJ: Hampton Press.

Brotchie John, Michael Batty, Peter Hall and Peter Newton, eds, 1991, *City of the 21st Century*, Cheshire: Longman.

Brotchie, John, Michael Batty, Ed Blakely, Peter Hall and Peter Newton, eds, 1995, *Cities in Competition: Productive and Sustainable Cities for the 21st Century*, Melbourne: Longman Australia.

Brown, Robert, 1991, *An Initial Study of the Global-Economy Urban Hierarchy*, Phoenix: Bronze Age Publishers.

Brownill, Sue, 1994, 'Selling the inner city: regeneration and place marketing in London's Docklands', in John R. Gold and Stephen V. Ward, eds, *Place Promotion: The Use of Publicity and Marketing to Sell Towns and Regions*, Chichester: John Wiley and Sons, pp. 133–152.

Bruinsma, Frank and Piet Rietveld, 1993, 'Urban agglomerations in European infrastructure networks', *Urban Studies*, **30** (6), pp. 919–934.

Brunn, Stanley D. and Thomas R. Leinbach, eds, 1991, *Collapsing Space and Time: Geographic Aspects of Communications and Information*, London: HarperCollins Academic.

Budd, Leslie, 1995, 'Globalisation, territory and strategic alliances in different financial centers', *Urban Studies*, **32** (2), pp. 345–360.

Budd, Leslie and Sam Whimster, eds, 1992, *Global Finance and Urban Living: A Study of Metropolitan Change*, London: Routledge.

Bukowczyk, John and Douglas Aikenhead, 1989, *Detroit Images: Photographs of the Renaissance City*, Detroit: Wayne State University.

Burgess, Jacquelin A., 1982, 'Selling places: environmental images for the executive', *Regional Studies*, **16** (1), pp. 1–17.

Burgess, Jacquelin and Peter Wood, 1988, 'Decoding Docklands: place advertising and the decision-making strategies of the small firm', in John Eyles and David M. Smith, eds, *Qualitative Methods in Human Geography*, Cambridge: Polity Press, pp. 94–117.

Business Week, 1994, 'Kansas City,' 26 December, pp. 33–54.

Button, K., ed., 1991, *Airline Deregulation: International Experiences*, New York: New York University Press.

Cairncross, Frances, 1997, *The Death of Distance: How the Communications Revolution Will Change Our Lives*, Boston: Harvard Business School Press.

Carey, John, ed., 1987, *Eye-Witness to History*, London: Faber and Faber.

Carnoy, Martin, 1993, 'Multinationals in a changing world economy: whither the nation-state?', in Martin Carnoy, Manuel Castells, Stephen Cohen and Fernando Cardoso, eds, *The New Global Economy in the Information Age: Reflections on Our Changing World*, University Park: Pennsylvania State University Press, pp. 45–96.

Cassidy, John, 1998, 'The triumphalist', *The New Yorker*, 6 July, pp. 54–60.

Castells, Manuel, 1989, *The Informational City: Information Technology, Economic Restructuring, and the Urban-regional Process*, Cambridge: Basil Blackwell.

Castells, Manuel, 1994, 'European cities, the informational society, and the global economy', *New Left Review*, **204**, pp. 18–32.

Castells, Manuel and Peter Hall, 1994, *Technopoles of the World: The Making of Twenty-First-Century Industrial Complexes*, London: Routledge.

Castles, Stephen and Mark J. Miller, 1993, *The Age of Migration: International Population Movements in the Modern World*, New York: Guilford Press.

Cattan, Nadine, 1995a, 'Attractivity and internationalisation of major European cities: the example of air traffic', *Urban Studies*, **32** (2), pp. 303–312.

Cattan, Nadine, 1995b, 'Barrier effects: the case of air and rail flows', *International Political Science Review*, **16**, pp. 237–248.

Chauncey, George, 1994, *Gay New York*, New York: HarperCollins.

Church, Andrew and Peter Reid, 1996, 'Urban power, international networks and competition: the example of cross-border cooperation', *Urban Studies*, **33** (8), pp. 1297–1319.

Clark, H., 1994, 'Taking up space: redefining political legitimacy in New York City', *Environment and Planning A*, **26**, pp. 937–955.

Clarke, Susan E. and Gary L. Gaile, 1998, *The Work of Cities*, Minneapolis: University of Minnesota Press.

Clusa, J., 1996, 'Barcelona: economic development 1970–1995', in N. Harris and I. Fabricius, eds, *Cities and Structural Adjustment*, London: University of California Press, pp. 102–116.

Coakley, Jerry, 1992, 'London as an international financial centre', in Leslie Budd and Sam Whimster, eds, *Global Finance and Urban Living: A Study of Metropolitan Change*, London: Routledge, pp. 52–72.

Cochrane, Allan, Jamie Peck and Adam Tickell, 1996, 'Manchester plays games: exploring the local politics of globalization', *Urban Studies*, **33** (8), pp. 1319–1336.

Code, William R., 1991, 'Information flows and the processes of attachment and projection: the case of financial intermediaries', in Stanley D. Brunn and Thomas R. Leinbach, eds, *Collapsing Space and Time: Geographic Aspects of Communications and Information*, London: HarperCollins Academic, pp. 111–131.

Coffey, William J., 1996, 'Forward and backward linkages of producer-services establishments: evidence from the Montreal metropolitan area', *Urban Geography*, **17** (7), pp. 604–632.

Cohen, Benjamin J., 1996, 'Phoenix risen: the resurrection of global finance', *World Politics*, **48**, pp. 268–296.

Cohen, R. B., 1981, 'The new international division of labor, multinational corporations and urban hierarchy', in Michael Dear and Allen Scott, eds, *Urbanization and Urban Planning in Capitalist Society*, London: Methuen, pp. 287–315.

Cooke, Philip and Frank Moulaert, eds, 1992, *Towards Global Localization: the Computing and Telecommunications Industries in Britain and France*, London: UCL Press.

Corbridge, Stuart, 1994, 'Bretton Woods revisited: hegemony, stability, and territory', *Environment and Planning A*, **26**, pp. 1829–1859.

Corbridge, Stuart, Nigel Thrift, and Ron Martin, eds, 1994, *Money, Power and Space*, Oxford: Blackwell.

Corey, Kenneth E., 1991, 'The role of information technology in the planning and development of Singapore', in Stanley D. Brunn and Thomas R. Leinbach, eds, *Collapsing Space and Time: Geographic Aspects of Communications and Information*, London: HarperCollins Academic, pp. 217–231.

Cornwell, Johnathon, 1997, 'The language man – David Crystal,' *The Weekend Australian*, 21 June, pp. 42–43.

Cosgrove, Denis, 1994, 'Contested global visions: one-world, whole-earth, and the Apollo space photographs', *Annals of the Association of American Geographers*, **84** (2), pp. 270–294.

Cox, Kevin R., 1993, 'The local and the global in the new urban politics: a critical view', *Environment and Planning D*, **11**, pp. 433–448.

Cox, Kevin R., 1995, 'Globalisation, competition and the politics of local economic development', *Urban Studies*, **32** (2), pp. 213–224.

Cox, Kevin R., ed., 1997, *Spaces of Globalization: Reasserting the Power of the Local*, New York: Guilford Press.

Cox, Kevin R. and Andrew E. Jonas, 1993, 'Urban development, collective consumption and the politics of metropolitan fragmentation', *Political Geography*, **12** (1), pp. 8–37.

Cox, Kevin R. and Andrew Mair, 1988, 'Locality and community in the politics of local economic development', *Annals of the Association of American Geographers*, **78** (2), pp. 307–325.

Cox, Kevin R. and Andrew Mair, 1989, 'Book review essay: urban growth machines and the politics of local economic development', *International Journal of Urban and Regional Research*, **13** (1), pp. 137–146.

Cox, Robert W., 1996, 'A perspective on globalization', in James H. Mittelman, ed., *Globalization: Critical Reflections*, Boulder: Lynne Rienner, pp. 21–30.

Crewe, L. and M. Lowe, 1995, 'Gap on the map? Towards a geography of consumption and identity', *Environment and Planning A*, **27** (12), pp. 1877–1898.

Crilley, Darrel, 1993, 'Architecture as advertising: constructing the image of redevelopment', in Gerry Kearns and Chris Philo, eds, *Selling Places: The City as Cultural Capital, Past and Present*, Oxford: Pergamon Press, pp. 231–252.

Cronon, W., 1991, *Nature's Metropolis: Chicago and The Great West*, New York: W. W Norton.

Crothers, T., 1995, 'The shakedown', *Sports Illustrated*, 19 June, pp. 78–82.

Crystal, David, 1997, *English as a Global Language*, Cambridge: Cambridge University Press.

Daniels, Peter W., 1991, 'Internationalization, telecommunications and metropolitan development: the role of producer services', in Stanley D. Brunn and Thomas R. Leinbach, eds, *Collapsing Space and Time: Geographic Aspects of Communications and Information*, London: HarperCollins Academic, pp. 149–169.

Daniels, Peter W., 1993, *Service Industries in the World Economy*, Oxford: Blackwell.

Daniels, Peter W., 1995, 'Services in a shrinking world', *Geography*, **80** (2), pp. 97–110.

Daniels, Peter W. and W. F. Lever, 1996, *The Global Economy in Transition*, Essex: Longman.

Dear, Michael and Steven Flusty, 1998, 'Postmodern urbanism', *Annals of the Association of American Geographers*, **88** (1), pp. 50–72.

DeLeon, Richard E., 1992, 'The urban antiregime: progressive politics in San Francisco', *Urban Affairs Quarterly*, **27** (4), pp. 555–579.

Dicken, Peter, 1998, *Global Shift: Transforming the World Economy* (3rd edition), New York: Guilford Press.

Dicken, Peter, Mats Forsgren and Anders Malmberg, 1994, 'The local embeddedness of transnational corporations', in Ash Amin and Nigel Thrift, eds, *Globalization, Institutions, and Regional Development in Europe*, Oxford: Oxford University Press, pp. 23–45.

Diehl, Paul F., ed., 1997, *The Politics of Global Governance: International Organizations in an Interdependent World*, Boulder: Lynne Rienner Publishers.

Dieleman, Frans and Chris Hamnett, 1994, 'Globalisation, regulation and the urban system:

editors' introduction to the special issue', *Urban Studies*, **31** (3), pp. 357–364.

Dieleman, Frans, Hugo Priemus and Wim Blauw, 1993, 'European cities: changing urban structures in a changing world', *Tijdschrift voor Economische en Sociale Geografie*, **84** (4), pp. 242–246.

Dissanayake, Wimal, 1996, 'Asian cinema and the American cultural imaginary', *Theory, Culture and Society*, **13** (4), pp. 109–122.

Dongailbo, 1998, 'IMF requests on the merge of two failing Korean banks', 16 February (http://www.dongailbo.co.kr) (in Korean).

Doro, Sue, 1992, *Blue Collar Goodbyes*, Watsonville: Papier-Mache Press.

Drache, Daniel and Meric S. Gertler, eds, 1991, *The New Era of Global Competition: State Policy and Market Power*, Montreal: McGill-Queen's University Press.

Drennan, Matthew P., 1992, 'Gateway cities: the metropolitan sources of US producer service exports', *Urban Studies*, **29** (2), pp. 217–235.

Drennan, Matthew, Emanuel Tobier and Jonathan Lewis, 1996, 'The interruption of income convergence and income growth in large cities in the 1980s', *Urban Studies*, **33** (1), pp. 63–82.

du Gay, Paul, 1996, *Consumption and Identity at Work*, London: Sage Publications.

Ducatel, Ken and Ian Miles, 1992, 'Internationalization of information technology services and public policy implications', *World Development*, **20** (12), pp. 1843–1857.

Dugger, Celia W., 1998, 'Wedding vows bind old world and new', *The New York Times* (http://www.nytimes.com/library/national/regional/072098 immigration.html)

Dunning, John H., 1993, *The Globalization of Business: the Challenge of the 1990s*, London: Routledge.

Dunning, John H. and Rajneesh Narula, eds, 1996, *Foreign Direct Investment and Governments: Catalysts for Economic Restructuring*, London: Routledge.

Eggers, William D., 1993, 'America's boldest mayors', *Policy Review*, Summer, pp. 67–74.

Eigen, Peter, 1996, 'Combating corruption around the world', *Journal of Democracy*, 7, pp. 158–160.

Environment and Planning A, 1996, 'On the nation-state, the global, and social science', **28** (11), pp. 1917–1995.

Esparza, A. and A. Krmenec, 1994, 'Producer services trade in city systems: evidence from Chicago', *Urban Studies*, **31** (1), pp. 29–46.

Fainstein, Susan S., 1991, 'Promoting economic development: urban planning in the United States and Great Britain', *Journal of the American Planning Association*, **57** (1), pp. 22–33.

Fainstein, Susan S., 1994, *The City Builders: Property, Politics, and Planning in London and New York*, Oxford: Blackwell.

Fainstein, Susan S., Ian Gordon and Michael Harloe, eds, 1992, *Divided Cities: New York and London in the Contemporary World*, Oxford: Blackwell.

Far Eastern Economic Review, 1995, 'Going global', 2 November, pp. 46–52.

Far Eastern Economic Review, 1996a, 'The post-national economy: goodbye widget, hello Nike', 29 August, p. 5.

Far Eastern Economic Review, 1996b, 'South Korea: trade and investment', 20 June, pp. 41–49.

Feagin, Joe R. and Michael Peter Smith, 1987, 'Cities and the new international division of labour: an overview', in Michael Peter Smith and Joe R. Feagin, eds, *The Capitalist City: Global Restructuring and Community Politics*, Oxford: Basil Blackwell, pp. 3–34.

Featherstone, Mike, 1990, 'Global culture: an introduction', *Theory, Culture and Society*, 7 (2/3), pp. 1–14.

Featherstone, Mike, 1993, 'Global and local cultures', in Jon Bird, Barry Curtis, Tim Putnam, George Robertson and Lisa Tickner, eds, *Mapping the Futures: Local Cultures, Global Change*, London: Routledge, pp. 169–187.

Featherstone, Mike, 1995, *Undoing Culture: Globalization, Postmodernism and Identity*, London: Sage Publications.

Financial Times, 1990, 'European finance and investment', 29 November, pp. 35 and 38.

Financial World, 1993, 'Memphis: target practice', 2 March, p. 51.

Financial World, 1993, 'Fairfax country,' 24 May, pp. 51–54.

Fincher, Ruth and Jane M. Jacobs, eds, 1998, *Cities of Difference*, New York: Guilford Press.

Findlay, A. M., F. L. N. Li, A. J. Jowett and R. Skeldon, 1996, 'Skilled international migration and the global city: a study of expatriates in Hong Kong', *Transactions of the Institute of British Geographers*, **21** (1), pp. 49–61.

Fleming, Douglas and Richard Roth, 1991, 'Place in advertising', *Geographical Review*, **81** (3), pp. 281–291.

Foner, Nancy, 1997, 'The immigrant family: cultural legacies and cultural changes', *International Migration Review*, **31** (4), pp. 961–974.

Forbes, 1994, 'Greater Milwaukee: the city that works for your business,' 19 December, pp. 51–30.

Forbes, 1997, 'Hot spots' (http://www.forbes.com/forbes/97/0922/6006278a.htm).

Forsström, Ake and Sten Lorentzon, 1991, 'Global development of communication: a frame for the pattern of localization in a small industrialized country', in Stanley D. Brunn and Thomas R. Leinbach, eds, *Collapsing Space and Time: Geographic Aspects of Communications and Information*, London: HarperCollins Academic, pp. 82–107.

Fortune, 1961, 1966, 1971, 1976, 1981, 1986, 'The 500 largest U.S. industrial corporations', July in 1961 and 1966; May in others.

Fortune, 1961, 1966, 1971, 1976, 1981, 1986, 'The largest industrial corporations outside the U.S.', August.

Fortune, 1976, 1981, 1986, 1991, 1994, 'The largest industrial companies in the world', August in 1976, 1981 and 1986; July in others.

Fortune, 1997, 1998, 'The global 500" (http://www.pathfinder.com/fortune/global500/).

Freeman, Gary P., 1998, 'The decline of sovereignty? Politics and immigration restriction in liberal states', in Christian Joppke, ed., *Challenge to the Nation-State: Immigration in Western Europe and the United States*, Oxford: Oxford University Press, pp. 86–108.

Friedman, Jonathan, 1990, 'Being in the world: globalization and localization', *Theory, Culture and Society*, 7 (2/3), pp. 311–328.

Friedmann, John, 1986, 'The world city hypothesis', *Development and Change*, 17, pp. 69–83.

Friedmann, John, 1995, 'Where we stand: a decade of world city research', in Paul L. Knox and Peter J. Taylor, eds, *World Cities in a World System*, Cambridge: Cambridge University Press, pp. 21–47.

Friedmann, John and Goets Wolff, 1982, 'World city formation: an agenda for research and action', *International Journal of Urban and Regional Research*, 6 (3), pp. 309–344.

Frost, Martin and Nigel Spence, 1993, 'Global city characteristics and central London's employment', *Urban Studies*, 30 (3), pp. 547–558.

Fujita, Kuniko, 1991, 'A world city and flexible specialization: restructuring of the Tokyo metropolis', *International Journal of Urban and Regional Research*, 15 (2), pp. 269–284.

Fuller, Stephen S., 1989, 'The internationalization of the Washington, D.C. area economy', in Richard V. Knight and Gary Gappert, eds, *Cities in a Global Society*, Newbury Park: Sage, pp. 108–119.

Fulton, William, 1997, *The Reluctant Metropolis: The Politics of Urban Growth in Los Angeles*, Point Arena, CA: Solano Press Books.

Gaffikin, Frank and Barney Warf, 1993, 'Urban policy and the post-Keynesian state in the United Kingdom and the United States', *International Journal of Urban and Regional Research*, 17 (1), pp. 67–84.

Ganz, Alexander and L. Francois Konga, 1989, 'Boston in the world economy', in Richard V. Knight and Gary Gappert, eds, *Cities in a Global Society*, Newbury Park: Sage, pp. 132–140.

Garcia, Linda, 1995, 'The globalization of telecommunications and information', in William J. Drake, ed., *The New Information Infrastructure: Strategies for U.S. Policy*, New York: Twentieth Century Fund Press, pp. 75–92.

Garcia, Soledad, 1996, 'Cities and citizenship', *International Journal of Urban and Regional Research*, 20 (1), pp. 7–21.

Geddes, Patrick, 1995, *Cities in Evolution*, London: Benn.

Gibbs, David and Brian Leach, 1994, 'Telematics in local economic development: the case of Manchester', *Tijdschrift voor Economische en Sociale Geografie*, 85 (3), pp. 209–223.

Gillespie, A. and H. Williams, 1988, 'Telecommunications and the reconstruction of regional comparative advantage', *Environment and Planning A*, 20, pp. 1311–1321.

Godfrey, B. and Y. Zhou, in press, 'World cities and the global urban hierarchy'.

Goetz, Andrew, 1992, 'Air passenger transportation and growth in the U.S. urban system 1950–1987', *Growth and Change*, 23, pp. 217–238.

Goetz, Andrew, 1993, 'Geographic patterns of air service frequencies and pricing at U.S. cities', *Journal of Transportation Research Forum*, 33, pp. 56–72.

Goetz, Andrew and C. Sutton, 1997, 'The geography of deregulation in the U.S. airline industry', *Annals of the Association of American Geographers*, 86, pp. 238–163.

Gold, John R., 1994, 'Locating the message: place promotion as image communication', in John R. Gold and Stephen V. Ward, eds, *Place Promotion: The Use of Publicity and Marketing to Sell Towns and Regions*, Chichester: John Wiley and Sons, pp. 19–38.

Gold, John R. and Stephen V. Ward, eds, 1994, *Place Promotion: The Use of Publicity and Marketing to Sell Towns and Regions*, Chichester: John Wiley and Sons.

Gomez, Maria V., 1998, 'Reflective images: the case of

urban regeneration in Glasgow and Bilbao', *International Journal of Urban and Regional Research*, **22** (1), pp. 106–121.

Goodman, John and Louis Pauly, 1993, 'The obsolescence of capital controls?: economic management in an age of global markets', *World Politics*, **46**, pp. 50–82.

Goodwin, Mark, 1993, 'The city as commodity: the contested spaces of urban development', in Gerry Kearns and Chris Philo, eds, *Selling Places: The City as Cultural Capital, Past and Present*, Oxford: Pergamon Press, pp. 145–162.

Gordon, David, 1988, 'The global economy: new edifice or crumbling foundations?', *New Left Review*, **168**, pp. 24–64.

Goss, J. D., 1993, 'Placing the market and marketing place: tourist advertising of the Hawaiian Islands, 1972–92', *Environment and Planning D: Society and Space*, **11**, pp. 663–688.

Graham, Daniel and Nigel Spence, 1995, 'Contemporary deindustrialisation and tertiarisation in the London economy', *Urban Studies*, **32** (6), pp. 885–911.

Graham, Daniel and Nigel Spence, 1997, 'Competition for metropolitan resources: the "crowding out" of London's manufacturing industry?', *Environment and Planning A*, **29** (3), pp. 459–484.

Graham, Stephen, 1994, 'Networking cities: telematics in urban policy – a critical review', *International Journal of Urban and Regional Research*, **18** (3), pp. 416–432.

Graham, Stephen, 1997, 'Cities in the real-time age: the paradigm challenge of telecommunications to the conception and planning of urban space', *Environment and Planning A*, **29** (2), pp. 105–127.

Graham, Stephen and Simon Marvin, 1996, *Telecommunications and the City: Electronic Spaces, Urban Places*, London: Routledge.

Gray, Matthew, 1997, 'Internet statistics: Web growth summary" (http://www.mit.edu/people/mkgray/net/web-growth-summary.html).

Griffiths, R., 1993, 'The politics of cultural policy in urban regeneration strategies', *Policy and Politics*, **21**, pp. 39–46.

Grosfoguel, Ramon, 1995, 'Global logics in the Caribbean city system: the case of Miami', in Paul L. Knox and Peter J. Taylor, eds, *World Cities in a World System*, Cambridge: Cambridge University Press, pp. 156–170.

Grube, John, 1997, ' "No More Shirt": the struggle for democratic gay space in Toronto,' in Gordon Ingram, Anne-Marie Bouthillette and Yolanda Retter, eds, *Queers in Space*, Seattle: Bay Press, pp. 127–145.

Haider, Donald, 1992, 'Place wars: new realities of the 1990s', *Economic Development Quarterly*, **6** (2), pp. 127–134.

Hajer, Maarten A., 1993, 'Rotterdam: re-designing the public domain', in Franco Bianchini and Michael Parkinson, eds, *Cultural Policy and Urban Regeneration: the West European Experience*, Manchester University Press, pp. 48–72.

Hall, Peter, 1984, *The World Cities*, London: Weidenfeld & Nicolson.

Hall, Peter, 1993, 'Forces shaping urban Europe', *Urban Studies*, **30** (6), pp. 883–898.

Hall, Peter, 1996, 'The global city', *International Social Science Journal*, **147**, pp. 15–24.

Hall, Peter and Sidney Tarrow, 1998, 'Globalization and area studies: when is too broad too narrow?', *The Chronicle of Higher Education*, 23 January, pp. B4–B5.

Hall, Stuart, 1997, 'The local and the global: globalization and ethnicity', in Anthony D. King, ed., *Culture, Globalization and the World-System: Contemporary Conditions for the Representation of Identity*, Minneapolis: University of Minnesota Press, pp. 19–40.

Hall, Tim and Phil Hubbard, 1996, 'The entrepreneurial city: new urban politics, new urban geographies?', *Progress in Human Geography*, **20** (2), pp. 153–174.

Hall, Tim and Phil Hubbard, eds, 1998, *The Entrepreneurial City: Geographies of Politics, Regime and Representation*, Chichester: John Wiley and Sons.

Hamnett, Chris, 1994, 'Social polarisation in global cities: theory and evidence', *Urban Studies*, **31** (3), pp. 401–424.

Hamnett, Chris, 1996, 'Why Sassen is wrong: a response to Burgers', *Urban Studies*, **33** (1), pp. 107–110.

Hanlon, P., 1996, *Global Airlines: Competition in a Transnational Industry*, Oxford: Butterworth-Heinemann.

Hannerz, Ulf, 1997, 'Scenarios for peripheral cultures', in Anthony D. King, ed., *Culture, Globalization and the World-System: Contemporary Conditions for the Representation of Identity*, Minneapolis: University of Minnesota Press, pp. 107–128.

Harding, Alan, 1994, 'Urban regimes and growth machines: toward a cross-national research agenda', *Urban Affairs Quarterly*, **29** (3), pp. 356–382.

Harvey, David, 1989, 'From managerialism to entrepreneurialism: the transformation in urban governance in late capitalism', *Geografiska Annaler*, **71B** (1), pp. 3–18.

Harvey, David, 1992, 'Social justice, postmodernism and the city', *International Journal of Urban and Regional Research*, **16** (4), pp. 588–601.

Harvey, David, 1995, 'Globalization in question', *Rethinking Marxism*, **8** (4), pp. 1–17.

Harvey, Thomas, 1996, 'Portland, Oregon: regional city in a global economy', *Urban Geography*, **17** (1), pp. 95–114.

Held, David, 1995, *Democracy and the Global Order: from the Modern State to Cosmopolitan Governance*, Stanford: Stanford University Press.

Hepworth, Mark, 1989, *Geography of the Information Economy*, London: Belhaven Press.

Hepworth, Mark, 1990, 'Planning for the information city: the challenge and response', *Urban Studies*, **27** (4), pp. 537–558.

Hepworth, Mark, 1991, 'Information technology and the global restructuring of capital markets', in Stanley D. Brunn and Thomas R. Leinbach, eds, *Collapsing Space and Time: Geographic Aspects of Communications and Information*, London: HarperCollins Academic, pp. 132–148.

Hepworth, Mark and Ken Ducatel, 1992, *Transport in the Information Age: Wheels and Wires*, London: Belhaven Press.

Herring, Richard J. and Robert E. Litan, 1995, *Financial Regulation in the Global Economy*, Washington, DC: The Brookings Institution.

Hewison, R., 1987, *The Heritage Industry*, London: Methuen.

Hicks, Donald A. and Steven R. Nivin, 1996, 'Global credentials, immigration, and metro-regional economics', *Urban Geography*, **17** (1), pp. 23–43.

Higgins, Ean, 1998, 'Barcelona after the Olympic Games,' *The Weekend Australian*, 4 July, pp. 20–23.

Hill, Dilys M., 1994, *Citizens and Cities: Urban Policy in the 1990s*, New York: Harvester Wheatsheaf.

Hirst, Paul and Grahame Thompson, 1992, 'The problem of "globalization": international economic relations, national economic management and the formation of trading blocs', *Economy and Society*, **21** (4), pp. 357–396.

Hirst, Paul and Grahame Thompson, 1996, *Globalization in Question: the International Economy and the Possibilities of Governance*, Cambridge: Blackwell.

Hitz, Hansruedi, Christian Schmid and Richard Wolff, 1994, 'Urbanization in Zürich: headquarter economy and city-belt', *Environment and Planning D*, **12** (2), pp. 167–185.

Holcomb, Briavel, 1993, 'Revisioning place: de- and re-constructing the image of the industrial city', in Gerry Kearns and Chris Philo, eds, *Selling Places: The City as Cultural Capital, Past and Present*, Oxford: Pergamon Press, pp. 133–143.

Holcomb, Briavel, 1994, 'City make-overs: marketing the post-industrial city', in John R. Gold and Stephen V. Ward, eds, *Place Promotion: The Use of Publicity and Marketing to Sell Towns and Regions*, Chichester: John Wiley and Sons, pp. 115–132.

Holt, Douglas B., 1998, 'Does cultural capital structure American consumption?', *Journal of Consumer Research*, **25** (1), pp. 1–25.

Howlett, Debbie, 1995, 'Monet is Chicago's stroke of fortune', *USA Today*, 12 July, p. A3.

Hubbard, Phil, 1996a, 'Re-imaging the city: the transformation of Birmingham's urban landscape', *Geography*, **81** (1), pp. 26–36.

Hubbard, Phil, 1996b, 'Urban design and city regeneration: social representations of entrepreneurial landscapes', *Urban Studies*, **33** (8), pp. 1441–1461.

Hymer, Stephen, 1972, 'The multinational corporation and the law of uneven development', in Jagdish Bhagwati, ed., *Economics and World Order from the 1970s to the 1990s*, London: Macmillan, pp. 113–135.

Imrie, Rob, Huw Thomas and Tim Marshall, 1995, 'Business organisations, local dependence and the politics of urban renewal in Britain', *Urban Studies*, **32** (1), pp. 31–46.

Imrie, Rob, Steven Pinch and Mark Boyle, 1996, 'Identities, citizenship and power in the cities', *Urban Studies*, **33** (8), pp. 1255–1261.

Ingram, Gordon, Anne-Marie Bouthillette and Yolanda Retter, eds., 1997, *Queers in Space*, Seattle: Bay Press.

Institute of International Education, 1980/81–1996/97, *Open Doors: Report on International Educational Exchange*, New York.

International Civil Aviation Organization (ICAO), 1985, 1990, 1992, 1996a, 1997, *On-Flight Origin and Destination*.

International Civil Aviation Organization (ICAO), 1996b, 'ICAO news release' (http://www.com.org/~icao/pio9608.htm).

International Department, The Bank of Korea, 1996, *Overseas Direct Investment Statistics Yearbook*, Seoul.

International Economic Policy Bureau, Ministry of Finance and Economy, Republic of Korea, 1996, *Trends in Foreign Investment and Technology Inducement*, October, Seoul.

International Monetary Fund, 1997, *Balance of Payments Statistics Yearbook*, Part 2, Washington, DC.

International Monetary Fund, 1997, *International Finance Statistics Yearbook*, Washington, DC.

Internet Movie Database, 1998, 'Business information for Titanic (1997)' (http://us.imdb.com/More?business+Titanic+(1997)).

Ivy, R. L., 1995, 'The restructuring of air transport linkages in the new Europe', *Professional Geographer*, **43**, pp. 280–288.

Ivy, R. L., T. J. Fik and E. J. Malecki, 1995, 'Changes in air service connectivity and employment', *Environment and Planning A*, **27**, pp. 165–179.

Jackson, Peter, 1995, 'Changing geographies of consumption', *Environment and Planning A*, **27** (12), pp. 1875–1876.

Jackson, Peter and Morris B. Holbrook, 1995, 'Multiple meanings: shopping and the cultural politics of identity', *Environment and Planning A*, **27** (12), pp. 1913–1930.

Jackson, Peter and Nigel Thrift, 1995, 'Geographies of consumption', in D. Miller, ed., *Acknowledging Consumption*, London: Routledge, pp. 204–237.

Jacobs, Jane, 1984, *Cities and the Wealth of Nations: Principles of Economic Life*, New York: Random House.

Janelle, Donald G., 1991, 'Global interdependence and its consequences', in Stanley D. Brunn and Thomas R. Leinbach, eds, *Collapsing Space and Time: Geographic Aspects of Communications and Information*, London: HarperCollins Academic, pp. 49–81.

Jenkins, Brian and Spyros A. Sofos, eds, 1996, *Nation and Identity in Contemporary Europe*, London: Routledge.

Jeong, Kuk-Hwan and John Leslie King, 1997, 'Korea's national information infrastructure: vision and issues', in Brian Kahin and Ernest Wilson, eds, *National Information Infrastructure Initiatives: Vision and Policy Design*, Cambridge, MA: MIT Press, pp. 112–149.

Jessop, Bob, 1994, 'The transition to post-Fordism and the shumpeterian workfare state', in R. Burrows and B. Loader, eds, *Towards a Post-Fordist Welfare State?*, London: Routledge, pp. 13–37.

John, Peter and Alistair Cole, 1998, 'Urban regimes and local governance in Britain and France', *Urban Affairs Review*, **33** (3), pp. 382–404.

Joppke, Christian, ed., 1998, *Challenge to the Nation-State: Immigration in Western Europe and the United States*, Oxford: Oxford University Press.

Judge, David, Gerry Stoker and Harold Wolman, eds, 1995, *Theories of Urban Politics*, London: Sage Publications.

Kachru, Braj B., 1990, *The Alchemy of English: The Spread, Functions, and Models of Non-native Englishes*, Chicago: University of Illinois Press.

Kaplan, David H. and Alex Schwartz, 1996, 'Minneapolis-St. Paul in the global economy', *Urban Geography*, **17** (1), pp. 44–59.

Kasaba, Resat, ed., 1991, *Cities in the World-System*, New York: Greenwood Press.

Kearns, Gerry and Chris Philo, eds, 1993, *Selling Places: The City as Cultural Capital, Past and Present*, Oxford: Pergamon Press.

Keeling, David, 1995, 'Transport and the world city paradigm', in Paul L. Knox and Peter J. Taylor, eds, *World Cities in a World System*, Cambridge: Cambridge University Press, pp. 115–131.

Keeling, David, 1996, *Buenos Aires: Global Dreams, Local Crises*, Chichester: John Wiley & Sons.

Keil, Roger and Klaus Ronneberger, 1994, 'Going up the country: internationalization and urbanization on Frankfurt's northern fringe', *Environment and Planning D*, **12** (2), pp. 137–166.

Kellermann, A., 1993, *Telecommunications and Geography*, London: Belhaven Press.

Kenen, Peter B., 1994, *Managing the World Economy: Fifty Years After Bretton Woods*, Washington, DC: Institute for International Economics.

Kevin Matthews and Artifice, Inc., 1998, 'Artifice great buildings online – Frank Gehry' (http://www.greatbuildings.com/gbc/architects/Frank_Gehry.html)

Kim, J., S. Rhee, J. Yu, K. Ku and J. Hong, 1989, *The Impact of the Seoul Olympic Games on National Development*, Seoul: Korea Development Institute (in Korean).

Kim, Joochul and Sang-Chuel Choe, 1997, *Seoul: the Making of a Metropolis*, Chichester: John Wiley and Sons.

Kim, Yeong-Hyun, 1997, 'Interpreting the Olympic landscape in Seoul: the politics of sports, spectacle and landscape', *Journal of the Korean Geographical Society*, **32** (3), pp. 387–402.

Kim, Yeong-Hyun, 1998, *Globalization, Urban Changes and Seoul's Dreams: a Global Perspective on Contemporary Seoul*, unpublished PhD dissertation, Department of Geography, Syracuse University.

King, Anthony D., 1990a, *Global Cities: Post-Imperialism and the Internationalization of London*, London: Routledge.

King, Anthony D., 1990b, 'Architecture, capital and

the globalization of culture', *Theory, Culture and Society*, **7** (2/3), pp. 397–411.

King, Anthony D., ed., 1996, *Re-Presenting the City: Ethnicity, Capital and Culture in the 21st-Century Metropolis*, New York: New York University Press.

King, Anthony D., ed., 1997, *Culture, Globalization and the World-System: Contemporary Conditions for the Representation of Identity*, Minneapolis: University of Minnesota Press.

Klak, T., 1994, 'Havana and Kingston: mass media images and empirical observations of two Caribbean cities in crisis', *Urban Geography*, **15**, pp. 318–44.

Knight, Richard V., 1995, 'Knowledge-based development: policy and planning implications for cities', *Urban Studies*, **32** (2), pp. 225–260.

Knight, Richard V. and Gary Gappert, eds, 1989, *Cities in a Global Society*, Newbury Park: Sage.

Knox, Paul L., 1995, 'World cities in a world system', in Paul L. Knox and Peter J. Taylor, eds, *World Cities in a World System*, Cambridge: Cambridge University Press, pp. 3–20.

Knox, Paul L., 1996, 'Globalization and urban change', *Urban Geography*, **17** (1), pp. 115–117.

Knox, Paul L. and Peter J. Taylor, eds, 1995, *World Cities in a World System*, Cambridge: Cambridge University Press.

Kofman, Bleonore, 1995, 'Citizenship for some but not for others: spaces of citizenship in contemporary Europe', *Political Geography*, **14** (2), pp. 121–137.

Konings, R., E. Louw and P. Rietveld, 1992, 'Transport infrastructure in the Randstad: an international perspective', *Tijdschrift voor Economische en Sociale Geografie*, **83**, pp. 263–277.

Korea National Tourism Organization, 1995, *Annual Statistical Report on Tourism*, Seoul.

Korea National Tourism Organization, 1997, *International Convention Industry in Korea*, Seoul.

Korea Network Information Center (KRNIC), 1998, 'Internet hosts, 1993–1998' (http://www.krnic.net/net/2_93_00/html).

Korean Federation of Small Business, 1996, 'The industrial training program for foreigners', 12 November.

Korea Telecom, 1997, *Statistical Yearbook of Telecommunications*, Seoul.

Koslowski, Rey, 1998, 'European Union migration regimes, established and emergent', in Christian Joppke, ed., *Challenge to the Nation-State: Immigration in Western Europe and the United States*, Oxford: Oxford University Press, pp. 153–188.

Kotler, Philip, Donald H. Haider and Irving Rein, 1993, *Marketing Places: Attracting Investment to Cities, States, and Nations*, New York: Free Press.

Kowarick, Lucio and Milton Campanario, 1986, 'Sao Paulo: the price of world city status,' *Development and Change*, **17**, pp. 159–174.

Kresl, Peter Karl, 1995, 'The determinants of urban competitiveness: a survey', in Peter Karl Kresl and Gary Gappert, eds, *North American Cities and the Global Economy: Challenges and Opportunities*, Thousand Oaks: Sage Publications, pp. 45–68.

Kristof, Nicholas, 1997a, 'South Korea moves closer to requesting I.M.F. aid', *The New York Times*, 21 November, p. D1.

Kristof, Nicholas, 1997b, 'Troubled economy stirs fears in South Korea', *The New York Times*, 10 November, p. A3.

Laferriere, Eric, 1994, 'Environmentalism and the global divide', *Environmental Politics*, **3**, pp. 91–113.

Langdale, John, 1985, 'Electronic funds transfer and the internationalisation of the banking and finance industry', *Geoforum*, **16** (1), pp. 1–13.

Langdale, John, 1989, 'The geography of international business telecommunications: the role of leased networks', *Annals of the Association of American Geographers*, **79** (4), pp. 501–522.

Langdale, John, 1991, 'Telecommunications and international transactions in information services', in Stanley D. Brunn and Thomas R. Leinbach, eds, *Collapsing Space and Time: Geographic Aspects of Communications and Information*, London: HarperCollins Academic, pp. 193–214.

Lanvin, Bruno, ed., 1993, *Trading in a New World Order: the Impact of Telecommunications and Data Services on International Trade in Services*, Boulder: Westview Press.

Larson, James F., 1995, *The Telecommunications Revolution in Korea*, Oxford: Oxford University Press.

Larson, James F. and Heung-Soo Park, 1993, *Global Television and the Politics of the Seoul Olympics*, Boulder: Westview Press.

Lash, S. and John Urry, 1994, *Economies of Signs and Spaces*, Beverly Hills, CA: Sage.

Law, Christopher M., 1992, 'Urban tourism and its contribution to economic regeneration', *Urban Studies*, **29** (3/4), pp. 599–618.

Lawless, Paul, 1994, 'Partnership in urban regeneration in the UK: the Sheffield central area study', *Urban Studies*, **31** (8), pp. 1303–1324.

Lee, Gun Young and Hyun Sik Kim, eds, 1995, *Cities and Nation: Planning Issues and Policies of Korea*,

Seoul: Korea Research Institute for Human Settlements.

Lee, J., L. Chen and S. Shaw, 1994, 'A method for the exploratory analysis of airline networks', *Professional Geographer*, **46**, pp. 468–477.

Lee, Roger and Ulrich Schmidt-Marwede, 1993, 'Interurban competition? financial centres and the geography of financial production', *International Journal of Urban and Regional Research*, **17** (4), pp. 492–515.

Leiken, Robert S., 1996–97, 'Controlling the global corruption epidemic', *Foreign Policy*, **105**, pp. 55–73.

Leislie, D. A., 1995, 'Global scan: the globalization of advertising agencies, concepts, and campaigns', *Economic Geography*, **71** (4), pp. 402–426.

Leitner, Helga, 1990, 'Cities in pursuit of economic growth: the local state as entrepreneur', *Political Geography Quarterly*, **9** (2), pp. 146–170.

Leontidou, Lila, 1996, 'Alternatives to modernism in (southern) urban theory: exploring in-between spaces', *International Journal of Urban and Regional Research*, **20** (1), pp. 178–195.

Lever, William F., 1993, 'Competition within the European urban system', *Urban Studies*, **30** (6), pp. 935–948.

Lever, William F., 1997, 'Delinking urban economies: the European experience', *Journal of Urban Affairs*, **19** (2), pp. 227–238.

Levine, Marc V., 1989, 'Urban redevelopment in a global economy: the cases of Montreal and Baltimore', in Richard V. Knight and Gary Gappert, eds, *Cities in a Global Society*, Newbury Park: Sage, pp. 141–152.

Lewis, Jim R., 1994, 'City challenge: involving the community in UK urban policy?', in Gerhard O. Braun, ed., *Managing and Marketing of Urban Development and Urban Life*, Berlin: Dietrich Reimer Verlag, pp. 367–378.

Lewis, Sinclair, 1922, *Babbitt*, New York: Harcourt, Brace and Company.

Ley, Robert and Pierre Poret, 1997, 'The new OECD members and liberalization', *The OECD Observer*, **205** (April/May), pp. 38–42.

Lipietz, Alain, 1993, 'The local and the global: regional individuality or interregionalism', *Transactions of Institute of British Geographers*, **18** (1), pp. 8–18.

Lo, F. C. and Y. M. Yeung, 1997, *Emerging World Cities in Pacific Asia*, Tokyo: United Nations University Press.

Loftman, Patrick and Brendan Nevin, 1996, 'Going for growth: prestige projects in three British cities', *Urban Studies*, **33** (6), pp. 991–1019.

Loftman, Patrick and Brendan Nevin, 1998, 'Pro-growth local economic development strategies: civic promotion and local needs in Britain's second city, 1981–1996', in Tim Hall and Phil Hubbard, eds, *The Entrepreneurial City: Geographies of Politics, Regime and Representation*, Chichester: John Wiley and Sons, pp. 129–148.

Logan, John R. and Harvey L. Molotch, 1987, *Urban Fortunes: The Political Economy of Place*, Berkeley: University of California Press.

Longcore, Travis R. and Peter W. Rees, 1996, 'Information technology and downtown restructuring: the case of New York City's financial districts', *Urban Geography*, **17** (4), pp. 354–372.

Lowe, Michelle, 1993, 'Local hero! An examination of the role of the regional entrepreneur in the regeneration of Britain's regions', in Gerry Kearns and Chris Philo, eds, *Selling Places: The City as Cultural Capital, Past and Present*, Oxford: Pergamon Press, pp. 211–230.

Lustiger-Thaler, Henri and Eric Shragge, 1998, 'The new urban left: parties without actors', *International Journal of Urban and Regional Research*, **22** (2), pp. 233–244.

Lyons, Donald and Scott Salmon, 1995, 'World cities, multinational corporations, and urban hierarchy: the case of the United States', in Paul L. Knox and Peter J. Taylor, eds, *World Cities in a World System*, Cambridge: Cambridge University Press, pp. 98–114.

Machimura, Takashi, 1992, 'The urban restructuring process in Tokyo in the 1980s: transforming Tokyo into a world city', *International Journal of Urban and Regional Research*, **16** (1), pp. 114–128.

Machimura, Takashi, 1998, 'Symbolic use of globalization in urban politics in Tokyo', *International Journal of Urban and Regional Research*, **22** (2), pp. 183–194.

Maltby, Richard, 1989, *Passing Parade: a History of Popular Culture in the Twentieth Century*, Oxford: Oxford University Press.

Mander, Jerry and Edward Goldsmith, eds, 1996, *The Case against the Global Economy and for a Turn toward the Local*, San Francisco: Sierra Club Books.

Marek, S. Makowski, 1992, 'Private networks directory', *Satellite Communications*, September.

Markusen, Ann and Vicky Gwiasda, 1994, 'Multipolarity and the layering of functions in world cities: New York city's struggle to stay on top', *International Journal of Urban and Regional Research*, **18** (2), pp. 167–193.

Martin, Ron, 1994, 'Stateless monies, global financial integration and national economic autonomy: the end of geography?', in Stuart Corbridge, Nigel Thrift and Ron Martin, eds, *Money, Power and Space*, Oxford: Blackwell, pp. 253–278.

Marx, Karl and Friedrich Engels, 1968 (first published 1872), 'Manifesto of the Communist Party', in *Marx and Engels: Selected Works*, London: Lawrence and Wishart.

Masai, Yasuo, 1989, 'Greater Tokyo as a global city', in Richard V. Knight and Gary Gappert, eds, *Cities in a Global Society*, Newbury Park: Sage, pp. 153–163.

Mayer, M., 1995, 'Urban governance in the post-Fordist city', in P. Healey, S. Cameron, S. Davoudi, S. Graham and A. Madani-Pour, eds, *Managing Cities: the New Urban Context*, New York: John Wiley & Sons, pp. 231–249.

McConville, Daniel J., 1993, 'Making it in Memphis', *Distribution*, November, pp. 50–58.

McDonald's Corporation, 1998, 'McDonald's around the world' (http://www.mcdonalds.com/ surfthe world/surf.html).

McGrew, Anthony, 1992, 'A global society?', in Stuart Hall, David Held and Tony McGrew, eds, *Modernity and Its Futures*, Cambridge: Polity Press, pp. 62–116.

McGuirk, P. M., H. P. M. Winchester and K.M. Dunn, 1996, 'Entrepreneurial approaches to urban decline: the Honeysuckle redevelopment in inner New Castle, New South Wales', *Environment and Planning A*, 28 (10), pp. 1815–1841.

Mead, Walter Russell, 1998, 'The new global economy takes your order,' *Mother Jones,* March/April (http://www.motherjones.com/mother_jones/ MA98/mead.html).

Meijer, Martine, 1993, 'Growth and decline of European cities: changing positions of cities in Europe', *Urban Studies*, 30 (6), pp. 981–990.

Mercer, K., 1991, 'Welcome to the jungle', in J. Rutherford, ed., *Identity: Community, Culture, Difference*, London: Lawrence and Wishart, pp. 43–71.

Meyer, David R., 1986, 'The world system of cities: relations between international financial metropolises and South American cities', *Social Forces*, 64 (3), pp. 553–581.

Meyer, David R., 1991a, 'Change in the world system of metropolises: the role of business intermediaries', *Urban Geography*, 12 (5), pp. 393–416.

Meyer, David R., 1991b, 'The formation of a global financial center: London and its intermediaries', in Resat Kasaba, ed., *Cities in the World-System*, New York: Greenwood Press, pp. 97–106.

Michael S. Monaco, 1998, Olympic Almanac (http://www.andrew.cmu.edu/~mmdg/Almanac/).

Miron, Louis F., 1992, 'Corporate ideology and the politics of entrepreneurism in New Orleans', *Antipode*, 24 (4), pp. 263–288.

Mitchell, Mark and Dave Russell, 1996, 'Immigration, citizenship and the nation-state in the New Europe', in Brian Jenkins and Spyros A. Sofos, eds, *Nation and Identity in Contemporary Europe*, London: Routledge, pp. 54–80.

Mittelman, James H., 1996, *Globalization: Critical Reflections*, Boulder: Lynne Rienner.

Mlinar, Zdravko, ed., 1992, *Globalization and Territorial Identities*, Aldershot: Avebury.

Mollenkopf, John and Manuel Castells, eds, 1991, *Dual City: Restructuring New York*, New York: Russell Sage Foundation.

Molotch, Harvey, 1976, 'The city as a growth machine: toward a political economy of place', *American Journal of Sociology*, 82 (2), pp. 309–332.

Mookerjee, Ajay and James Cash, 1990, *Global Electronic Wholesale Banking*, London: Graham and Trotman.

Moricz, Zoltan and Laurence Murphy, 1997, 'Space traders: reregulation, property companies and Auckland's office market, 1975–94', *International Journal of Urban and Regional Research*, 21 (2), pp. 165–179.

Morita, Kiriro and Saskia Sassen, 1994, 'The new illegal immigration in Japan, 1980–1992', *International Migration Review*, 28 (1), pp. 153–163.

Morley, David and Kevin Robins, 1995, *Spaces of Identity: Global Media, Electronic Landscapes, and Cultural Boundaries*, London: Routledge.

Moss, Mitchell L., 1987, 'Telecommunications, world cities, and urban policy', *Urban Studies*, 24 (6), pp. 534–546.

Moss, Mitchell L., 1988, 'Telecommunications: shaping the future', in George Sternlieb and James W. Hughes, eds, *America's New Market Geography: Nation, Region and Metropolis*, New Brunswick, NJ: Rutgers University Press, pp. 255–275.

Moss, Mitchell L., 1991, 'The information city in the global economy', in John Brotchie, Michael Batty, Peter Hall and Peter Newton, eds, *Cities of the 21st Century: New Technologies and Spatial Systems*, New York: Longman Cheshire, pp. 181–189.

Moulaert, Frank and Faridah Djellal, 1995, 'Information technology consultancy firms: economies of agglomeration from a wide-area perspective', *Urban Studies*, 32 (1), pp. 105–122.

Moulaert, Frank and Arie Shachar, 1995, 'Special issue: cities, enterprises and society at the eve of the

21st century: introduction', *Urban Studies*, **32** (2), pp. 205–212.

Murie, Alan and Sako Musterd, 1996, 'Social segregation, housing tenure and social change in Dutch cities in the late 1980s', *Urban Studies*, **33** (3), pp. 495–516.

Murray, M., 1995, 'Correction at Cabrini-Gree; a socio-spatial exercise of power', *Environment and Planning D*, **13**, pp. 311–327.

Naisbitt, John, 1994, *Global Paradox: the Bigger the World Economy, the More Powerful Its Smallest Players*, New York: William Morrow and Company.

Neill, William, 1995a, 'Lipstick on the Gorilla: the failure of image-led planning in Coleman Young's Detroit', *International Journal of Urban and Regional Research*, **19** (4), pp. 639–653.

Neill, William, 1995b, 'Promoting the city: image, reality and racism in Detroit', in William Neill, Diana Fitzsimons and Brendan Murtagh, eds, *Reimaging the Pariah City: Urban Development in Belfast and Detroit*, Aldershot: Avebury, pp. 113–161.

Network Wizards, 1998, 'Internet domain survey: number of Internet hosts' (http://www.nw.com/zone/host-count-history).

Newman, Peter and Andy Thornley, 1997, 'Fragmentation and centralisation in the governance of London: influencing the urban policy and planning agenda', *Urban Studies*, **34** (7), pp. 967–988.

Nijman, Jan, 1996, 'Breaking the rules: Miami in the urban hierarchy', *Urban Geography*, **17** (1), pp. 5–22.

Nijman, Jan, 1997, 'Globalization to a Latin beat: the Miami growth machine', *The Annals of the American Academy of Political and Social Science*, **551**, pp. 164–177.

Noyelle, Thierry, 1989, 'New York's competitiveness', in Thierry Noyelle, ed., *New York's Financial Markets: The Challenges of Globalization*, Boulder: Westview Press, pp. 91–114.

Noyelle, Thierry, ed., 1989, *New York's Financial Markets: The Challenges of Globalization*, Boulder: Westview Press.

Ó Tuathail, Gearóid and Timothy W. Luke, 1994, 'Present at the (dis)integration: deterritorialization and reterritorialization in the new wor(l)d order', *Annals of the Association of American Geographers*, **84** (3), pp. 381–398.

O'Brien, Richard, 1992, *Global Financial Integration: The End of Geography*, New York: Royal Institute of International Affairs.

O'Connor, Justin, 1998, 'Popular culture, cultural intermediaries and urban regeneration', in Tim Hall and Phil Hubbard, eds, *The Entrepreneurial City: Geographies of Politics, Regime and Representation*, Chichester: John Wiley and Sons, pp. 225–240.

O'Connor, Justin and Derek Wynne, eds, 1996, *From the Margins to the Centre: Cultural Production and Consumption in the Post-industrial City*, Brookfield: Arena.

O'Connor, Kevin and Ann Scott, 1992, 'Airline services and metropolitan areas in the Asia-Pacific region 1970–1990', *Review of Urban and Regional Development Studies*, **4**, pp. 240–253.

Ohmae, Kenichi, 1995, *The End of the Nation State: the Rise of Regional Economies*, New York: Free Press.

Oum, T., A. Zhang and Y. Zhang, 1995, 'Airline network rivalry', *Canadian Journal of Economics*, **28**, pp. 836–857.

Paddison, Ronan, 1993, 'City marketing, image reconstruction and urban regeneration', *Urban Studies*, **30** (2), pp. 339–350.

Painter, Joe, 1998, 'Entrepreneurs are made, not born: learning and urban regimes in the production of entrepreneurial cities', in Tim Hall and Phil Hubbard, eds, *The Entrepreneurial City: Geographies of Politics, Regime and Representation*, Chichester: John Wiley and Sons, pp. 259–274.

Parkinson, Michael and Franco Bianchini, 1993, 'Liverpool: a tale of missed opportunities', in Franco Bianchini and Michael Parkinson, eds, *Cultural Policy and Urban Regeneration: the West European Experience*, Manchester University Press, pp. 155–177.

Passell, Peter, 1997, 'One world, one economy, one big problem with currencies', *The New York Times*, 20 November, p. D2.

Peck, Jamie, 1995, 'Moving and shaking: business elites, state localism and urban privatism', *Progress in Human Geography*, **19** (1), pp. 16–46.

Peck, Jamie and Adam Tickell, 1994, 'Searching for a new institutional fix: the after-Fordist crisis and the global–local disorder', in Ash Amin, ed., *Post-Fordism: A Reader*, Oxford: Blackwell, pp. 280–315.

Peck, Jamie and Adam Tickell, 1995, 'Business goes local: dissecting the "business agenda" in Manchester', *International Journal of Urban and Regional Research*, **19** (1), pp. 55–78.

Persky, Joseph and Wim Wiewel, 1994, 'The growing localness of the global city', *Economic Geography*, **70** (2), pp. 129–143.

Peterson, Paul E., 1981, *City Limits*, Chicago: University of Chicago Press.

Podmore, Julie, 1998, '(Re)reading the "loft living" habitus in Montreal's inner city', *International Journal of Urban and Regional Research*, **22** (2), pp. 283–302.

Policy Analysis Computing and Information Facility In Commerce, University of British Columbia, 1998, 'Pacific exchange service' (http://pacific.commerce.ubc.ca/xr/data.html).

Pollack, Andrew, 1997a, 'South Korea halts activity at 9 banks', *The New York Times*, 3 December, pp. D1 and D4.

Pollack, Andrew, 1997b, 'Package of loans worth $55 billion is set for Korea', *The New York Times*, 4 December, pp. A1 and D6.

Pollack, Andrew, 1997c, 'Frugal Koreans rush to rescue their rapidly sinking economy', *The New York Times*, 18 December, pp. A1 and D9.

Pollack, Andrew and Peter Passell, 1997, 'South Korea currency soars as trading restrictions end', *The New York Times*, 16 December, pp. A1 and D11.

Pratt, A. C., 1997, 'The cultural industries production system: a case study of employment change in Britain, 1984–91', *Environment and Planning A*, **29** (11), pp. 1953–1974.

Price, David G. and Alastair M. Blair, 1989, *The Changing Geography of the Service Sector*, London: Belhaven Press.

Pryke, Michael, 1991, 'An international city going "global": spatial change in the City of London', *Environment and Planning D*, **9** (2), pp. 197–222.

Pryke, Michael, 1994a, 'Urbanizing capitals: towards an integration of time, space and economic calculation', in Stuart Corbridge, Nigel Thrift and Ron Martin, eds, *Money, Power and Space*, Oxford: Blackwell, pp. 218–252.

Pryke, Michael, 1994b, 'Looking back on the space of a boom: (re)developing spatial matrices in the City of London', *Environment and Planning A*, **26** (2), pp. 235–264.

Pryke, Michael and Roger Lee, 1995, 'Place your bets: towards an understanding of globalization, socio-financial engineering and competition within a financial centre', *Urban Studies*, **32** (2), pp. 329–344.

Puhl, Elizabeth, 1998, *The Monano Terrace Convention Center*, unpublished Master thesis, Department of Geography, Syracuse University.

Quilley, Stephen, 1997, 'Constructing Manchester's "New Urban Village": gay space in the entrepreneurial city' in Gordon Ingram, Anne-Marie Bouthillette and Yolanda Retter, eds, *Queers in Space*, Seattle: Bay Press, pp. 275–292.

Raco, Mike, 1997, 'Business associations and the politics of urban renewal: the case of the Lower Don Valley, Sheffield', *Urban Studies*, **34** (3), pp. 383–402.

Rainey, J., 1995, 'L.A. hopes big sell will polish city's image', *Los Angeles Times*. 13 March, pp. A3 and A18.

Razin, Eran and Ivan Light, 1998, 'Ethnic entrepreneurs in America's largest metropolitan areas', *Urban Affairs Review*, **33** (3), pp. 332–360.

Reed, Howard Curtis, 1989, 'Financial center hegemony, interest rates, and the global political economy', in Yoon S. Park and Musa Essayyad, eds, *International Banking and Financial Centers*, Boston: Kluwer Academic Publishers, pp. 247–268.

Reich, Robert, 1991, *The Work of Nations*, New York: Alfred A. Knopf.

Rieff, David, 1991, *Los Angeles: Capital of the Third World*, New York: Simon & Schuster.

Rimmer, Peter J., 1986, 'Japan's world cities: Tokyo, Osaka, Nagoya or Tokaido Megapolis?', *Development and Change*, **17**, pp. 121–157.

Ritzer, George, 1993, *The McDonaldization of Society*, London: Sage.

Roberts, Susan, 1994, 'Fictitious capital, fictitious spaces: the geography of offshore financial flows', in Stuart Corbridge, Nigel Thrift and Ron Martin, eds, *Money, Power and Space*, Oxford: Blackwell, pp. 91–115.

Roberts, Susan M. and Richard H. Schein, 1993, 'The entrepreneurial city: fabricating urban development in Syracuse', *Professional Geographer*, **45** (1), pp. 21–32.

Robertson, Roland, 1990, 'Mapping the global condition: globalization as the central concept', *Theory, Culture and Society*, **7** (2/3), pp. 15–30.

Robertson, Roland, 1992, *Globalization: Social Theory and Global Culture*, London: Sage.

Robertson, Roland, 1997, 'Social theory, cultural relativity and the problem of globality,' in Anthony D. King, ed., *Culture, Globalization and the World-System: Contemporary Conditions for the Representation of Identity*, Minneapolis: University of Minnesota Press, pp. 69–90.

Robinson, William I., 1996, *Promoting Polyarchy: Globalization, US Intervention, and Hegemony*, Cambridge: Cambridge University Press.

Rose, N., 1990, *Governing the Soul: the Shaping of the Private Self*, London: Routledge.

Rosenblat, Celine and Denise Pumain, 1993, 'The location of multinational firms in the European urban system', *Urban Studies*, **30** (10), pp. 1691–1709.

Rosentraub, Mark S., David Swindell, Michael Przybylski and Daniel R. Mullins, 1994, 'Sport and downtown development strategy: if you build it, will jobs come?', *Journal of Urban Affairs*, **16** (3), pp. 221–239.

Roulac, S., 1993, 'Place wars and the Olympic Games,' *The Futurist*, November/December, pp. 18–19.

Rowe, David, 1996, 'The global love-match: sport and television', *Media, Culture and Society*, **18** (4), pp. 565–582.

Rowe, David, Geoffrey Lawrence, Toby Miller and Jim McKay, 1994, 'Global sport? Core concern and peripheral vision', *Media, Culture and Society*, **16** (4), pp. 661–676.

Rubalcaba-Bermejo, Luis and Juan R. Cuadrado-Roura, 1995, 'Urban hierarchies and territorial competition in Europe: exploring the role of fairs and exhibitions', *Urban Studies*, **32** (2), pp. 379–400.

Rutheiser, Charles, 1996, *Imagineering Atlanta: the Politics of Place in the City of Dreams*, London: Verso.

Ryan, K. Bruce, 1990, 'The official image of Australia', in Leo Zonn, ed., *Place Images in Media: Portrayal, Experience, and Meaning*, Savage: Rowman & Littlefield, pp. 135–158.

Sachs, Jeffrey, 1998, 'The IMF and the Asian flu', *The American Prospect*, March–April, pp. 16–21.

Sadler, David, 1993, 'Place-marketing, competitive places and the construction of hegemony in Britain in the 1980s', in Gerry Kearns and Chris Philo, eds, *Selling Places: The City as Cultural Capital, Past and Present*, Oxford: Pergamon Press, pp. 175–192.

Sadler, D., 1997, 'The global music business as an information industry: reinterpreting economies of culture', *Environment and Planning A*, **29** (11), pp. 1919–1936.

Salomon, Alan, 1995, 'Memphis distributes word on its benefits', *Advertising Age*, 2 October, p. 14.

Sampson, Anthony, 1991, 'Global money', *New Perspectives Quarterly*, **8** (4), pp. 64–66.

Sassen, Saskia, 1990, 'Finance and business services in New York: international linkages and domestic effects', *International Social Science Journal*, **42**, pp. 287–306.

Sassen, Saskia, 1991, *The Global City: New York, London, Tokyo*, Princeton: Princeton University Press.

Sassen, Saskia, 1994, *Cities in a World Economy*, Thousand Oaks, CA: Fine Forge Press.

Sassen, Saskia, 1995, 'On concentration and centrality in the global city', in Paul L. Knox and Peter J. Taylor, eds, *World Cities in a World System*, Cambridge: Cambridge University Press, pp. 63–75.

Sassen, Saskia, 1996, *Losing Control?: Sovereignty in an Age of Globalization*, New York: Columbia University Press.

Sassen, Saskia, 1998, 'The *de facto* transnationalizing of immigration policy', in Christian Joppke, ed., *Challenge to the Nation-State: Immigration in Western Europe and the United States*, Oxford: Oxford University Press, pp. 49–85.

Sassen-Koob, Saskia, 1986, 'New York city: economic restructuring and immigration', *Development and Change*, **17**, pp. 85–119.

Scanlon, Rosemary, 1989, 'New York City as global capital in the 1980s', in Richard V. Knight and Gary Gappert, eds, *Cities in a Global Society*, Newbury Park: Sage, pp. 83–95.

Schneider, Mark and Paul Teske, 1993, 'The progrowth entrepreneur in local government', *Urban Affairs Quarterly*, **29** (2), pp. 316–327.

Schuck, Peter H., 1998, 'The re-evaluation of American citizenship', in Christian Joppke, ed., *Challenge to the Nation-State: Immigration in Western Europe and the United States*, Oxford: Oxford University Press, pp. 191–230.

Sciorra, Joseph, 1996, 'Return to the future: Puerto Rican vernacular architecture in New York City', in Anthony D. King, ed., *Re-Presenting the City: Ethnicity, Capital and Culture in the 21st-Century Metropolis*, New York: New York University Press, pp. 60–92.

Scott, Allen J., 1997, 'The cultural economy of cities', *International Journal of Urban and Regional Research*, **21** (2), pp. 323–339.

Scott, Allen J. and Edward W. Soja, eds, 1998, *The City: Los Angeles and Urban Theory at the End of the Twentieth Century*, Berkeley: University of California Press.

Shachar, Arie, 1994, 'Randstad Holland: a "world city"?', *Urban Studies*, **31** (3), pp. 381–400.

Shachar, Arie, 1995, 'World cities in the making: the European context', in Peter Karl Kresl and Gary Gappert, eds, *North American Cities and the Global Economy: Challenges and Opportunities*, Thousand Oaks: Sage Publications, pp. 150–170.

Shefter, Martin, ed., 1993, *Capital of the American Century: the National and International Influence of New York City*, New York: Russell Sage Foundation.

Short, John Rennie, 1996, *The Urban Order: an Introduction to Cities, Culture, and Power*, Cambridge: Blackwell.

Short, John R., in press, 'Urban imagineers: boosterism and the representation of cities', in A. Jonas and D.

Wilson, eds, *The Urban Growth Machine*, Albany: SUNY Press.

Short, John Rennie and Yeong-Hyun Kim, 1998, 'Urban crises/urban representations: selling the city in difficult times', in Tim Hall and Phillip Hubbard, eds, *The Entrepreneurial City: Geographies of Politics, Regime and Representation*, Chichester: John Wiley and Sons, pp. 55–75.

Short, J. R., L. M. Benton, W. B. Luce and J. Walton, 1993, 'Reconstructing the image of an industrial city', *Annals of the Association of American Geographers*, **83** (2), pp. 207–224.

Short, J. R., Y. Kim, M. Kuus and H. Wells, 1996, 'The dirty little secret of world cities research: data problems in comparative analysis', *International Journal of Urban and Regional Research*, **20** (4), pp. 697–717.

Simon, David, 1995, 'The world city hypothesis: reflections from the periphery', in Paul L. Knox and Peter J. Taylor, eds, *World Cities in a World System*, Cambridge: Cambridge University Press, pp. 132–155.

Singh, Daljit and Reza Y. Siregar, eds, 1995, *ASEAN and Korea: Emerging Issues in Trade and Investment Relations*, Singapore: Institute of Southeast Asian Studies.

Sklair, Leslie, 1996, 'Conceptualizing and researching the transnational capitalist class in Australia', *Australian and New Zealand Journal of Sociology*, **32** (2), pp. 1–19.

Sklair, Leslie, 1998, 'Globalization and the corporations: the case of the California *Fortune* global 500', *International Journal of Urban and Regional Research*, **22** (2), pp. 195–215.

SLOOC (The Seoul Olympic Organizing Committee), 1989, *Games of the 24th Olympiad Seoul 1988: the Abridged Official Report*, Seoul.

Smith, David A. and Michael Timberlake, 1995a, 'Conceptualising and mapping the structure of the world system's city system', *Urban Studies*, **32** (2), pp. 287–302.

Smith, David A. and Michael Timberlake, 1995b, 'Cities in global matrices: toward mapping the world-system's city system', in Paul L. Knox and Peter J. Taylor, eds, *World Cities in a World System*, Cambridge: Cambridge University Press, pp. 79–97.

Smyth, Hedley, 1994, *Marketing the City: The Role of Flagship Developments in Urban Regeneration*, London: E & FN Spon.

Soja, Edward W., 1987, 'Economic restructuring and the internationalization of the Los Angeles region', in Joe R. Feagin and Michael Peter Smith, eds, *The Capitalist City: Global Restructuring and Community Politics*, Oxford: Basil Blackwell, pp. 178–198.

Soja, Edward W., 1989, *Postmodern Geographies: the Reassertion of Space in Critical Social Theory*, London: Verso.

Soja, Edward W., 1996, *The City: Los Angeles and Urban Theory at the End of the Twentieth Century*, Berkeley: University of California Press.

Sontag, Deborah, 1998, 'A Mexican town that transcends all borders' (http://www.nytimes.com/library/national/regional/072198immigration.html).

Sontag, Deborah and Celia W. Dugger, 1998, 'The new immigrant tide: a shuttle between worlds', *The New York Times*, 19 July, pp. 1, 28–30.

Staal, van der P., 1994, 'Communication media in Japan: economic and regional aspects', *Telecommunications Policy*, **18** (1), pp. 32–50.

Stevenson, Nick, 1997, 'Globalization, national cultures and cultural citizenship', *The Sociological Quarterly*, **38** (1), pp. 41–66.

Stewart, Cameron, 1998, 'Atlanta after the Olympic Games,' *The Weekend Australian*, 4 July, pp. 24–26.

Stone, Clarence, 1989, *Regime Politics: Governing Atlanta 1946–1988*, Lawrence: University of Kansas Press.

Storper, Michael, 1992, 'The limits to globalization: technology districts and international trade', *Economic Geography*, **68** (1), pp. 60–93.

Strange, Susan, 1994, 'From Bretton Woods to the casino economy', in Stuart Corbridge, Nigel Thrift and Ron Martin, eds, *Money, Power and Space*, Oxford: Blackwell, pp. 49–62.

Strange, Susan, 1995, 'The limits of politics', *Government and Opposition*, **30** (3), pp. 291–311.

Swyngedouw, Erik, 1996, 'Reconstructing citizenship, the re-scaling of the state and the new authoritarianism: closing the Belgian mines', *Urban Studies*, **33** (8), pp. 1499–1521.

Swyngedouw, Erik, 1997, 'Neither global nor local: "globalization" and the politics of scale', in Kevin R. Cox, ed., *Spaces of Globalization: Reasserting the Power of the Local*, New York: Guilford Press, pp. 137–166.

Tardanico, Richard and Mario Lungo, 1995, 'Local dimensions of global restructuring: changing labour-market contours in urban Costa Rica', *International Journal of Urban and Regional Research*, **19** (2), pp. 223–249.

Taylor, Peter J., 1996, 'Embedded statism and the social sciences: opening up to new spaces',

Environment and Planning A, **28** (11), pp. 1917–1928.

The Banker, 1970, 1976, 1981, 1986, 1991, 1997, 'Top banks', June in 1970, 1976 and 1981; July in others.

The Economist, 1992, 'Financial centres', 27 June , pp. S3–26.

The Economist, 1995, 'The Brave Marketeer of Milwaukee', 17 June, p. 34.

The Economist, 1997a, 'The Asian miracle: is it over?', 1 March, pp. 23–25.

The Economist, 1997b, 'Trade in Services', 3 May, p. 99.

The Economist, 1997c, 'Who will listen to Mr. Clean?', 2 August, p. 52.

The Economist, 1997d, 'Telecommunications: a connected world', 13 September, pp. S1–34.

The Economist, 1997e, 'The future of the state', 20 September, pp. S1–48.

The Economist, 1997f, 'South Korea: the end of the miracle', 29 November, pp. 21–23.

The Economist, 1997g, 'Asia's economic crisis: how far is down?', 15 November, pp. 19–21.

The Ministry of Government Administration and Home Affairs, Republic of Korea, 1985–1994, *National Statistical Yearbook*, Seoul.

The Ministry of Information and Communications, Republic of Korea, 1996, *News on the 2002 World Cup*, Seoul.

The Ministry of Information and Communication, Republic of Korea, 1997, *1997 White Paper* (http://www.mic.go.kr/english/paper/contents.htm).

The Ministry of Justice, Republic of Korea, 1996, 'Foreign workforce', August.

The Petroleum Economist Ltd and TeleGeography, Inc., 1996, *Telecommunications Map of the World*, New York: ING Barings.

The Seoul Metropolitan Government, 1995a, *A Proposal of Seoul Urban Planning toward 2011*, Seoul (in Korean).

The Seoul Metropolitan Government, 1995b, *A Survey of Seoul's Industrial Structure*, Seoul (in Korean).

The Seoul Metropolitan Government, 1995c, *Seoul 1394–1994: the Sixth Centennial toward a New Birth*, Seoul (in Korean).

The Seoul Metropolitan Government, 1995d, *A Report of the Sister-City Project*, Seoul. (in Korean)

The World Bank, 1997, *The State in a Changing World*, World Development Report 1997, Washington, DC: Oxford University Press.

Thompson, E. P., 1968, *The Making of the English Working Class*, London: Penguin.

Thrift, Nigel, 1989, 'Geography of international economic disorder', in R. J. Johnston and P. J. Taylor, eds, *A World in Crisis?: Geographical Perspectives* (2nd ed.), Oxford: Basil Blackwell, pp. 16–78.

Thrift, Nigel, 1994a, 'Globalisation, regulation, urbanization: the case of the Netherlands', *Urban Studies*, **31** (3), pp. 365–380.

Thrift, Nigel, 1994b, 'On the social and cultural determinants of international financial centres: the case of the City of London', in Stuart Corbridge, Nigel Thrift and Ron Martin, eds, *Money, Power and Space*, Oxford: Blackwell, pp. 327–355.

Thrift, Nigel and Andrew Leyshon, 1994, 'A phantom state? The de-traditionalization of money, the international financial system and international financial centres', *Political Geography*, **13** (4), pp. 299–327.

Todd, Graham, 1995, '"Going global" in the semi-periphery: world cities as political projects: the case of Toronto', in Paul L. Knox and Peter J. Taylor, eds, *World Cities in a World System*, Cambridge: Cambridge University Press, pp. 192–212.

Tödtling, Franz, 1994, 'The uneven landscape of innovation poles: local embeddedness and global networks', in Ash Amin and Nigel Thrift, eds, *Globalization, Institutions, and Regional Development in Europe*, Oxford: Oxford University Press, pp. 68–92.

Tokyo Stock Exchange, 1982, 1991, 1997, *Fact Book*.

Tomlinson, Alan, 1996, 'Olympic spectacle: opening ceremonies and some paradoxes of globalization', *Media, Culture and Society*, **18** (4), pp. 583–602.

Union of International Associations, 1996/1997, *Yearbook of International Organizations*, Vol. 1 *Organization Descriptions and Cross-references*, Munchen: K. G. Saur.

USA Today, 1995, 'Financial district still a magnet', 7 December, p. 12B.

van Elteren, Mel, 1996, 'Conceptualizing the impacts of US popular culture globally', *Journal of Popular Culture*, **30** (1), pp. 47–89.

Walker, Richard, 1996, 'Another round of globalization in San Francisco', *Urban Geography*, **17** (1), pp. 60–94.

Wallerstein, Immanuel, 1997, 'The national and the universal: can there be such a thing as world cultures', in Anthony D. King, ed., *Culture, Globalization and the World-System: Contemporary Conditions for the Representation of Identity*, Minneapolis: University of Minnesota Press, pp. 91–106.

Walter, Ingo, 1988, *Global Competition in Financial Services: Market Structure, Protection, and Trade Liberalization*, Cambridge: An American Enterprise Institute/Ballinger Publication.

Ward, Peter M., 1995, 'The successful management and administration of world cities: mission impossible?', in Paul L. Knox and Peter J. Taylor, eds, *World Cities in a World System*, Cambridge: Cambridge University Press, pp. 298–314.

Ward, Stephen V., 1990, 'Local industrial promotion and development policies, 1899–1940', *Local Economy*, 5, pp. 100–118.

Ward, Stephen V., 1994, 'Time and place: key themes in place promotion in the USA, Canada and Britain since 1870', in John R. Gold and Stephen V. Ward, eds, *Place Promotion: The Use of Publicity and Marketing to Sell Towns and Regions*, Chichester: John Wiley and Sons, pp. 53–74.

Ward, Stephen V., 1996, 'Rereading urban regime theory: a sympathetic critique', *Geoforum*, 27, pp. 427–438.

Ward, Stephen V. and John R. Gold, 1994, 'Introduction', in John R. Gold and Stephen V. Ward, eds, *Place Promotion: The Use of Publicity and Marketing to Sell Towns and Regions*, Chichester: John Wiley and Sons, pp. 1–18.

Warf, Barney, 1989, 'Telecommunications and the globalization of financial services', *Professional Geographer*, 41 (3), pp. 257–271.

Warf, Barney, 1991, 'The internationalization of New York services', in P. W. Daniels, ed., *Services and Metropolitan Development: International Perspectives*, London: Routledge, pp. 245–264.

Warf, Barney, 1995, 'Telecommunications and the changing geographies of knowledge transmission in the late 20th century', *Urban Studies*, 32 (2), pp. 361–378.

Warf, Barney, 1996, 'International engineering services 1982–92', *Environment and Planning A*, 28, pp. 667–686.

Warf, Barney and Rodney Erickson, 1996, 'Introduction: globalization and the U.S. city system', *Urban Geography*, 17 (1), pp. 1–4.

Wark, McKenzie, 1994, *Virtual Geography: Living with Global Media Events*, Bloomington: Indiana University Press.

Waters, Malcolm, 1995, *Globalization*, London: Routledge.

Watson, S., 1991, 'Gilding the smokestacks: the new symbolic representations of deindustrialised regions', *Environment and Planning D*, 9 (1), pp. 59–70.

Weiss, Thomas G. and Leon Gordenker, eds, 1996, *NGOs, the UN, and Global Governance*, Boulder: Lynne Rienner Publishers.

Wernick, Andrew, 1991, *Promotional Culture: Advertising, Ideology and Symbolic Expression*, London: Sage.

Wetherall, M. and L. Potter, 1992, *Mapping the Language of Racism*, New York: Columbia University Press.

Who Owns Whom: a Directory of Parent, Associate and Subsidiary Companies, 1996, UK edition, London: O. W. Roskill and Co.

Williams, R., 1980, 'Advertising: magic system', in R. Williams, *Problems in Materialism and Culture*, London: Verso, pp. 170–195.

Willis, John, 1994–1996, *Screen World*, New York: Applause.

Wilson, D., 1989, 'Local state dynamics and gentrification in Indianapolis', *Urban Geography*, 10, pp. 19–40.

Wilson, D., 1995, 'Progress Report: Urban conflict politics and the poststructuralist gaze', *Urban Geography*, 16, pp. 134–147.

Wilson, Helen, 1996, 'What is an Olympic city? Visions of Sydney 2000', *Media, Culture and Society*, 18 (4), pp. 603–618.

Wilson, Patricia A., 1995, 'Embracing locality in local economic development', *Urban Studies*, 32 (4/5), pp. 645–658.

Wilson, Thomas M. and Hastings Donnan, eds, 1998, *Border Identities: Nation and State at International Frontiers*, Cambridge: Cambridge Unviersity Press.

Wulf, Steve, 1995, 'Bad bounces for the N.F.L.', *Time*, 11 December, pp. 64–65.

Wynne, Derek and Justin O'Connor, 1998, 'Consumption and the postmodern city', *Urban Studies*, 35 (5/6), pp. 841–864.

Yeung, Yue-man, 1996, 'An Asian perspective on the global city', *International Social Science Journal*, 147, pp. 25–32.

Young, Craig and Jonathan Lever, 1997, 'Place promotion, economic location and the consumption of city image', *Tijdschrift voor Economische en Sociale Geografie*, 88 (4), pp. 332–341.

Zelenko, Laura, 1992, 'Mid-size cities blitz NFL for new franchise', *American Demographics*, 14 (1), pp. 9–10.

Zelinsky, Wilbur, 1991, 'The twinning of the world: sister cities in geographic and historical perspective', *Annals of the Association of American Geographers*, 81 (1), pp. 1–31.

Zincone, G., 1989, '1988 welcome to Seoul', *Abitare*, 273, pp. 218–227.

Zukin, Sharon, 1982, *Loft Living: Culture and Capital*

in Urban Change, New Brunswick, NJ: Rutgers University Press.

Zukin, Sharon, 1991, *Landscapes of Power: from Detroit to Disney World*, Berkeley: University of California Press.

Zukin, Sharon, 1992, 'The city as a landscape of power: London and New York as global financial capitals', in Leslie Budd and Sam Whimster, eds, *Global Finance and Urban Living: a Study of Metropolitan Change*, London: Routledge, pp. 195–223.

Zukin, Sharon, 1995, *The Cultures of Cities*, Cambridge, MA: Blackwell.

Zukin, Sharon, Robert Baskerville, Miriam Greenberg, Courtney Guthreau, Jean Halley, Mark Halling, Kristin Lawler, Ron Nerio, Rebecca Stack, Alex Vitale and Betsy Wissinger, 1998, 'From Coney Island to Las Vegas in the urban imaginary: discursive practices of growth and decline', *Urban Affairs Review*, **33** (5), pp. 627–654.

Index

INTRODUCTION.

The index covers Chapters 1 to 10. Index entries are to page numbers. Alphabetical arrangement is word-by-word, where a group of letters followed by a space is filed before the same group of letters followed by a letter, e.g. 'global management' will appear before 'globalism'. Initial articles, conjunctions and prepositions are ignored in determining filing order.